Botana, Sainz (Eds.)
Climate Change and Mycotoxins

Also of interest

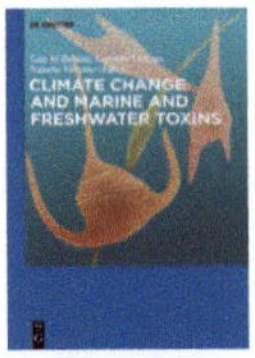

Climate Change and Marine and Freshwater Toxins
Botana, Louzao and Vilariño *(Eds.), 2015*
ISBN 978-3-11-033303-9, e-ISBN 978-3-11-033359-6

Chemistry of the Climate System
Detlev Möller, 2014
ISBN 978-3-11-033080-9, e-ISBN 978-3-11-033194-3

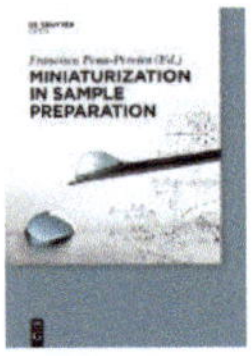

Miniaturization in Sample Preparation
Pena Pereira (Ed.); 2014
ISBN 978-3-11-041017-4, e-ISBN 978-3-11-041018-1

Introduction to Modern Instrumentation:
For Hydraulics and Environmental Sciences
Guaraglia and Pousa; 2014
ISBN 978-3-11-040171-4, e-ISBN 978-3-11-040172-1

Freshwater Fungi and Fungal-like Organisms
Jones, Hyde and Pang (Eds.), 2014
ISBN 978-3-11-033345-9, e-ISBN 978-3-11-033348-0

Reviews on Environmental Health
Carpenter, David O. and Sly, Peter (Editors-in-Chief)
ISSN 0048-7554, e-ISSN 2191-0308

Climate Change and Mycotoxins

Edited by
Luis M. Botana and María J. Sainz

DE GRUYTER

Editors
Prof. Luis M. Botana
Universidad de Santiago de Compostela
Facultad de Veterinaria
Departamento de Farmacología
27002 Lugo, Spain
Luis.botana@usc.es

Prof. María J. Sainz
Universidad de Santiago de Compostela
Facultad de Veterinaria
Departamento de Producción Vegetal
27002 Lugo, Spain
mj.sainz@usc.es

ISBN 978-3-11-033305-3
e-ISBN (PDF) 978-3-11- 033361-9
e-ISBN (EPUB) 978-3-11- 039015-5
Set-ISBN 978-3-11- 033362-6

Library of Congress Cataloging-in-Publication Data
A CIP catalog record for this book has been applied for at the Library of Congress.

Bibliographic information published by the Deutsche Nationalbibliothek
The Deutsche Nationalbibliothek lists this publication in the Deutsche Nationalbibliografie; detailed bibliographic data are available on the Internet at http://dnb.dnb.de.

Cover image: Dangdumrong/iStock/thinkstock
Typesetting: PTP-Berlin, Protago-TEX-Production GmbH, Berlin
Printing and binding: CPI books GmbH, Leck
♾ Printed on acid-free paper
Printed in Germany

www.degruyter.com

Preface

Climate change is expected to strongly affect agricultural productivity, food safety, and food security worldwide in the coming decades, due to higher temperatures, changes in water supply and availability, year-to-year climate variability, and extreme weather events. The impact on food and feed crops is expected to manifest not only in changes to plant physiology, yield, and quality, but also in incidence and severity of pests and diseases. Of particular concern are the potential changes in geographical distribution, natural inoculum, activity, and biological fitness of some of those fungal species which cause disease in staple food crops, particularly cereals, or grow saprophytically on stored plant products and produce mycotoxins.

Mycotoxins are secondary metabolites produced by filamentous fungi either pre- or postharvest and which can contaminate agricultural food and feed products and have detrimental effects on human and animal health. Although awareness of the fact that mycotoxins are ubiquitous in food and feed products is not very high, they are deeply ingrained in human culture, as their presence has been associated with deaths and serious feed and food poisoning throughout history. Those from *Claviceps* have even been an inspiration to artists, for example in paintings (e.g. by Peter Brueghel) and sculptures.

Nowadays, more than 100 countries have regulations specifying maximum tolerable levels for the most toxic and/or abundant mycotoxins, mainly in human food. Criteria for limits have not been harmonized worldwide, however. Much of the research on mycotoxins, including recent research on the effects of climate change on mycotoxins, has focused on the *Aspergillus*, *Fusarium*, and *Penicillium* species, as they are the major mycotoxin-producing fungi in field crops and stored products in the world.

When we decided to edit a book on climate change and mycotoxins, we were aware that, although it is relatively easy to identify climate change from the study of climatological records and the development of predictive models, it is not so easy to clearly define a link between climate change and expected mycotoxin risks in the different geographical zones of the world. In this book, the authors wrote outstanding chapters on the subject by reviewing the effects of changes in temperature, water availability, and CO_2 on the biodiversity, plasticity, occurrence, and mycotoxin profile of the most prevalent mycotoxigenic fungi on cereals in different regions, as there is no historical record to compare the amounts of toxins now and a century or more ago. The book provides information on the trends in mycotoxin occurrence in agricultural commodities over the last ten years. Toxins have been discovered by modern science, and therefore their presence, structure, or levels in food have only been known recently. The science of mycotoxins has significantly advanced in recent years as a result of the implementation of sensitive analytical detection technology, such as mass spectrometry, and the availability of the first certified standards; but improved standardized multitoxin detection analysis and further certified reference materials will have to be developed to support current and future mycotoxin regulation for the protection of human and animal health. They will also be essential to track any shift in the mycotoxin levels within the food chain worldwide in a climate change scenario.

This book intends to highlight the importance of the study of climate change impacts on mycotoxigenic fungi and their mycotoxins in food and feed crops in order to guarantee safe and sufficient food and feed in the future. The book is thus suitable for mycologists, mycotoxicologists, pathologists, epidemiologists, toxicologists, physicians, veterinarians, nutritionists, the food and feed industries, legislators, analytical chemists, microbiologists, biologists, or students of these fields.

As always, a book of this kind is unthinkable without the valuable work of the authors of the individual chapters. We wish to thank them, not only for their time, effort, and generosity in contributing their experience and expertise to this book, but also for their commitment to a difficult project, as is indeed the effect of climate change on food security.

The editors

Contents

List of contributing authors

Amparo Alfonso
Department of Pharmacology
Faculty of Veterinary
University of Santiago de Compostela
Campus s/n, 27002 Lugo, Spain
Chapter 8

José-Miguel Barea
Departamento de Microbiología del Suelo
y Sistemas Simbióticos
Estación Experimental del Zaidín (CSIC)
Profesor Albareda 1
18008 Granada, Spain
josemiguel.barea@eez.csic.es
Chapter 3

Eva M. Binder
Erber AG
Industriestraße 21
3130 Herzogenburg, Austria
eva.binder@erber-group.net
Chapter 7

Luis M. Botana (Ed.)
Department of Pharmacology
Faculty of Veterinary
University of Santiago de Compostela
Campus s/n, 27002 Lugo, Spain
luis.botana@usc.es
Chapter 8

Christian Joseph R. Cumagun
Crop Protection Cluster
College of Agriculture
University of the Philippines Los Baños
College, Laguna, Philippines 4031
Chapter 1

X. Li
College of Plant Protection
Hunan Agricultural University
Changsha, China
and
Department of Plant Pathology and Microbiology
Iowa State University
Ames, IA 50011, USA
linuslee@iastate.edu
Chapter 2

Yin-Jung Liu
Erber AG
Industriestraße 21
3130 Herzogenburg, Austria
Karin Nahrer
Biomin Holding GmbH
Industriestraße 21
3130 Herzogenburg, Austria
Chapter 7

Antonio F. Logrieco
Institute of Sciences of Food Production
Via Amendola 122/O
70126 Bari, Italy
antonio.logrieco@ispa.cnr.it
Chapter 5

Naresh Magan
Applied Mycology Group
Cranfield Soil and AgriFood Institute
Cranfield University, Cranfield
Bedford MK43 0AL, U.K.
Chapter 4

Angel Medina
Applied Mycology Group
Cranfield Soil and AgriFood Institute
Cranfield University, Cranfield
Bedford MK43 0AL, U.K.
a.medinavaya@cranfield.ac.uk
Chapter 4

Antonio Moretti
Institute of Sciences of Food Production
Via Amendola 122/O
70126 Bari, Italy
Chapter 5

Ireneo B. Pangga
Crop Protection Cluster
College of Agriculture
University of the Philippines Los Baños
College, Laguna, Philippines 4031
ibpangga@yahoo.com
Chapter 1

M. P. Kovalsky Paris
Biomin Holding GmbH
Industriestraße 21
3130 Herzogenburg, Austria
Chapter 7

Trond Rafoss
Norwegian Institute of Agricultural and
Environmental Research
Plant Health and Plant Protection Division
1430 Ås, Norway
Chapter 6

Alicia Rodriguez
Applied Mycology Group
Cranfield Soil and AgriFood Institute
Cranfield University, Cranfield
Bedford MK43 0AL, U.K.
Chapter 4

María J. Sainz (Ed.)
Department of Plant Production
University of Santiago de Compostela
Campus s/n, 27002 Lugo, Spain
mjsainz@usc.es
Chapter 8

Arnold R. Salvacion
Department of Community and Environmental
Resource Planning
College of Human Ecology
University of the Philippines Los Baños
College, Laguna, Philippines 4031
Chapter 1

Leif Sundheim
Norwegian Institute of Agricultural and
Environmental Research
Plant Health and Plant Protection Division
1430 Ås, Norway
leif.sundheim@bioforsk.no
Chapter 6

X. B. Yang
Department of Plant Pathology and Microbiology
Iowa State University
Ames, IA 50011, USA
xbyang@iastate.edu
Chapter 2

Ireneo B. Pangga, Arnold R. Salvacion, and Christian Joseph R. Cumagun

1 Climate change and plant diseases caused by mycotoxigenic fungi: implications for food security

1.1 Introduction

The contribution of Working Group I to the International Panel on Climate Change (IPCC) 5th Assessment Report (AR5) reaffirmed that the warming of the climate system is unequivocal and the observed changes are unprecedented over decades to millennia. Compared to the 4th Assessment Report (AR4), improved climate models and longer and more detailed observations showed that human influence on the climate system is clearly indicated by increasing greenhouse gas concentrations in the atmosphere and observed global warming. The atmospheric concentration of carbon dioxide (CO_2) of 391 ppm in 2011 has increased by 40 % since pre-industrial times due to emissions from fossil fuel combustion and changes in land use. The total change in energy fluxes caused by natural and anthropogenic drivers for 2011 relative to 1750 (radiative forcing) is positive, leading to energy uptake by the climate system with the largest contribution by CO_2 [1].

Representative Concentration Pathways (RCP), a new set of climate change scenarios, were identified by their approximate total radiative forcing (watts (W) per m^2) in the year 2100 relative to 1750: mitigation scenario with very low radiative forcing level (RCP2.6), two stabilization scenarios (RCP 4.5 and 6), and very high greenhouse emission scenario (RCP8.5). Climate model simulations make high confidence projections that the global temperature changes will likely exceed 1.5 °C at the end of the 21st century, relative to 1850–1990 for all RCP scenarios except RCP2.6 with 1 °C mean change. The global mean surface temperature change from 2016–2035 relative to 1986–2005 will likely be in the range of 0.3–0.7 °C (medium confidence). As the mean global temperature increases, it is virtually certain (99–100 % probability) that more frequent hot and fewer cold extremes in temperature will occur and that heat waves will very likely (90–100 % probability) occur with higher frequency and duration [1].

Changes in the global water cycle in response to global warming in the 21st century will not be uniform. The contrast between wet/dry regions and seasons will increase except for some regions. Extreme precipitation over most of the mid-latitude land masses and wet tropical regions will very likely become more intense and frequent at the end of the 21st century. In many mid-latitude and subtropical dry regions, mean precipitation will likely (66–100 % probability) decrease. Due to the increase in atmospheric moisture, the variability in El-Niño-Southern Oscillation (ENSO)-related precipitation in regional scales and monsoon precipitation will likely intensify [1].

The contribution of Working Group II (WG II) to IPCC AR5 assessed a substantially larger knowledge base of relevant scientific, technical and socioeconomic literature compared to past reports. This facilitated a comprehensive assessment of impacts, adaptation, and vulnerability of climate change risks in human and natural systems. WG II AR5 found that a high likelihood that all aspects of food security will potentially be affected by climate change, from food production to food access, utilization, and price stability [2].

Climate change poses a considerable challenge to food security as the global demand for food will increase to feed 9.2 billion people in 2050 [3, 4]. Without adaptation a negative impact of climate change on food production is projected for ≥ 2 °C temperature increase above late 20th century levels for wheat, rice, and maize in tropical and temperate areas, but impacts will vary across crops, regions, and adaptation scenarios [2]. Expanded geographic ranges and altered dynamics of insect pests and diseases due to climate change can exacerbate the reductions in crop yields [5–7]. Plant disease is a major constraint in food production, as direct yield losses due to pathogens cause 16 % overall loss in agricultural productivity [8], but the indirect harmful effects on food quality and safety are very serious in many crops and environments worldwide [9].

This chapter aims to discuss the effects of climate change on mycotoxigenic fungi on the three major food crops vital to food security such as wheat, rice, and maize. Maize, wheat, and rice provide 30 % of the food calories for 4.5 billion people in 100 developing countries [10]. Mycotoxin contamination has become one of the most important and challenging problems facing plant pathologists today [11, 12]. For example, in Fusarium head blight of wheat, environmental conditions such as rainfall and temperature are the dominant factor associated with disease infection [13, 14]. Knowledge of population genetics and epidemiology are the keys to Fusarium head blight management [15, 16]. The effects of climate change variables on epidemics in major food crops caused by pre- and post-harvest mycotoxigenic fungi will be analyzed in terms of plant disease epidemiology and population genetics.

1.1.1 Mycotoxigenic fungi and food security

Significant crop losses in major food crops due to mycotoxigenic fungi remain a major hindrance in achieving food security. Fusarium head blight of wheat and barley is considered a re-emerging disease due to historical and recent epidemics worldwide [17, 18]. In the USA, a direct economic loss of US $ 2.491 billion was observed due to Fusarium head blight from 1993–2001 [19]. In 2003, another Fusarium head blight epidemic in southeastern USA caused a pre-milling economic loss of over US $ 13.6 million [20].

The production of mycotoxins by fungi in major food crops is another important factor which has a severe impact on food security. It has been estimated that 25 % of the world food crops are affected by mycotoxins [21]. The three most important genera of mycotoxigenic fungi are *Aspergillus*, *Fusarium*, and *Penicillium*, producing the following classes of mycotoxins: aflatoxin (*Aspergillus*), ochratoxin (*Aspergillus* and *Penicillium*), trichothecenes and fumonisins (*Fusarium*) [22]. The potential annual cost of food and feed contamination in the US with aflatoxins, fumonisins, and deoxynivalenol (DON) was estimated at US $ 946 million [23]. Annual average losses of US $ 163 million to aflatoxins and US $ 40 million to fumonisins were estimated for US maize [24]. In 2010, 10 % of the Kenyan maize harvest was contaminated with aflatoxins resulting in an economic loss of approximately US $ 100 million [25]. In 2012, Aspergillus and Fusarium ear rot caused yield losses of 94.7 and 80.8 million bushels in top corn producing US states and Ontario, Canada with 18 % loss from mycotoxin contamination [26].

Mycotoxin contamination of the food chain is a food safety risk globally, leading to human health threats because it can cause mycotoxicosis, resulting in acute or chronic disease episodes [22]. The mycotoxin hazard can be exacerbated by food insecurity because mycotoxin-contaminated food can be consumed rather than discarded and malnutrition enhances the susceptibility to lower mycotoxin levels [27]. A fitting example of the adverse effect of mycotoxin contamination on human health is the 2004 aflatoxin contamination in Kenya where 125 people died [28]. In Thailand, Indonesia, and the Philippines the total annual social cost of aflatoxins in maize was AUS $ 319 million in 1991 [29].

1.1.2 Climate change and food security

Food production is an important aspect of food security, and it needs to be increased by 60 % by 2050 under current food consumption trends, assuming no significant reduction in food waste [30]. Climate change is projected to negatively affect food production as local temperature increases by 2 °C or more above late 20th century levels, but some individual locations will have positive impact (medium confidence). Process-based models and regional statistical analysis showed negative impacts of temperature above 30–34 °C on crop yields depending on the crop and region [5]. Based on statistical crop models and climate projections for 2030 from 20 general circulation models, the production of wheat in South Asia, rice in Southeast Asia, and maize in Southern Africa will suffer negative impact in the absence of adaptation [31].

Climate change could slow down the progress towards food security. Climate change can have a range of direct and indirect effects on all four dimensions of food security: food availability, access, utilization, and stability [2, 32]. The direct impacts of climate change on food availability will occur throughout the food chain but will be greatest for agriculture considering its climate sensitivity and key role in food supply.

Indirect impacts of climate change on nutrition, health, livelihoods, and poverty will be more complex and highly differentiated. The impacts of climate change on food systems are expected to be widespread, complex, geographically and temporally variable, and greatly influenced by socio-economic conditions [33]. The overall impact of climate change on food security is considered more complex and potentially greater than the projected impact on agricultural productivity [5]. Food security is diminished when food systems are stressed. Climate change effects on the food system will vary between regions [34] and food inequalities will increase from local to global levels due to spatial variation in climate change effects [32].

The impacts of climate change on food security will be worst in countries already suffering high levels of hunger [32]. Sub-Saharan Africa had the highest proportion of food insecure people with an estimated regional average of 20 % of the population is undernourished in 2010–2012, while the largest number is in South Asia with 300 million undernourished [30]. The risks of food insecurity and breakdown of food systems are linked to global warming, drought, flooding, and variable and extreme precipitation particularly for poorer populations in urban and rural settings of Africa, Asia, and Central and South America. Based on IPCC WG II AR5, reduced crop productivity associated with heat and drought stress will have strong adverse effects on food security while increased pest and disease damage and flood impacts on food system are projected in Africa. Increased risk of drought-related water and food shortage causing malnutrition is projected for Asia whereas decreased food production and food quality are projected for Central America [2].

Climate change can destabilize the food system resulting in high and volatile food prices [4, 35]. Increased temperature and altered precipitation without CO_2 effects will contribute to increased global food prices by 2050 from 3–84 % (medium confidence). Any negative impact of climate change on global crop yields is expected to lead to increases in international food prices and the proportion of population which is food-insecure [5].

1.1.3 Climate change effects on plant diseases and food security

Food security is defined as the access to sufficient, safe, and nutritious food for dietary needs and food preferences [36]. One major factor contributing to low productivity is crop loss due to plant health problems, which is poorly recognized as an important driver of food security [37]. Nevertheless, plant diseases have enormous impacts on food security as exemplified by the Irish potato famine in 1845 due to the potato late blight epidemic caused by *Phytophthora infestans*, which resulted in the death of 1 million people and the emigration of 1.5 million to mainland US [38]. A further example is the great Bengal famine of 1943 due to the rice brown spot epidemic caused by *Helminthosporium oryzae* in India, resulting in the starvation of 2 million people [39]. During these two plant disease epidemics, the weather conditions were very conducive for severe infection. One current example of a transboundary plant disease which can have a severe impact on global food production is the black stem rust of wheat caused by *Puccinia graminis tritici* race Ug99, a virulent strain which has already spread from Africa to the Middle East and is threatening wheat production in South Asia [40].

Around 10–16 % of the global harvest is estimated to result in financial losses of US $ 220 billion annually due to plant diseases [8, 41]. The worldwide actual losses to plant pathogens in major food crops from 2001–2003 were 10.2, 10.8, and 8.5 % for wheat, rice, and maize, respectively [8]. However, the limited quantitative data on crop losses hinders estimation of crop losses with significant effects on food security [37, 42]. According to Zeigler and Savary [43], plant diseases are key yield reducers. Any factor which reduces actual yield will impede increase in rice productivity and high elasticity in rice production-consumption; factors needed to prevent spikes in cereal prices which can lead to food crises.

1.2 Effects of climate change on plant diseases caused by mycotoxigenic fungi

The ability to predict the effects of climate change on plant pathogens and subsequently on yield is limited due to lack of data because studies have focused on individual diseases rather than a complete set [2, 9]. Studies showed that climate change can have positive, negative or neutral effects on individual pathosystems due to the specific nature of the host-pathogen interaction [44]. Indeed, the interaction of direct effects of climate change factors and indirect effects of global change factors contribute to pathosystem complexity under climate change, thereby hindering attempts to generalize pathogen responses [9, 45–47]. A recent analysis indicated that accelerated evolution and the changing geographic distribution of plant diseases are the main effects of climate change on plant pathogens [48]. Climate is a potent selective force in natural populations [49, 50] and can alter selection in host-parasite interactions [51].

1.2.1 Epidemiology and resistance

Climate change can alter stages and rates of pathogen development and modify host resistance and host-pathogen interactions [6, 44]. The assessment of climate change effects on fungal colonization and growth, and on mycotoxin production needs to be based on individual pathogens due to the different optimum conditions of temperature and water activity for growth and mycotoxin production [52]. The two most important factors affecting mycotoxigenic fungi are water availability or moisture and temperature [53, 54].

Tab. 1.1: A summary of recent findings on the influence of elevated CO_2 on disease intensity and mycotoxin and fitness of mycotoxigenic fungi.

Disease	Pathogen	Crop	Disease/Mycotoxin at elevated CO_2	Fitness/Aggressiveness	Reference
Crown rot	*Fusarium pseudo-graminearum*	Wheat	Increased fungal biomass and stem browning; Stem browning increased by 68 % without irrigation in partially resistant cv. 249 and not influenced by irrigation in susceptible cv. Tamaroi	No difference in saprophytic fitness between CO_2 levels	Melloy et al. [63]
Crown rot	*Fusarium pseudo-graminearum*; *Fusarium culmorum*	Wheat	Disease incidence increased with cropping cycles in susceptible cv. Tamaroi but not in resistant cv. 249. Resistance was induced in susceptible cv. Tamaroi. Reduced deoxynivalenol in grains	Aggressiveness increased by up to 110 % after 5 cropping cycles at both CO_2 levels. *F. pseudograminearum* increased frequency on susceptible cv. Tamaroi and decreased frequency on resistant cv. 249.	Khudhair et al. [61]
Ear rot	*Fusarium verticillioides*	Maize	Enhanced susceptibility; Reduced fumonisin levels		Vaughan et al. [62]

1.2.1.1 Elevated CO_2

There is limited knowledge about the effects of elevated CO_2 on mycotoxigenic fungi especially in the field. Previous *in vitro* studies have shown that mycotoxigenic fungi can tolerate concentrations of CO_2 which are much higher than the concentration projected under climate change scenarios [55]. Furthermore, fungal species growth and mycotoxin production interact with other factors such as moisture, temperature, and nutrients [56]. Based on the study of Samapundo et al. [57], 10 % CO_2 completely inhibited fumonisin B1 production by *Fusarium verticillioides*, but 40 % CO_2 is needed to inhibit fumonisin B1 production of *F. proliferatum* at 0.984 water activity (a_w). A modified atmosphere containing 20 % CO_2 generally inhibited mold growth, while 20–60 % CO_2 significantly reduced mycotoxin production by *Fusarium*, *Aspergillus*, and *Penicillium* spp. [58]. Regardless of moisture and temperature, latent periods preceding fungal growth were significantly increased by > 5 % CO_2, while sporulation was unaffected by increased CO_2 [56]. In free air carbon enrichment (FACE) studies which provide a more realistic climate change assessment of plant and disease interactions under natural field conditions [59], elevated CO_2 altered plant growth and development and increased disease levels of necrotrophic pathogens [60]. However, the effect of elevated CO_2 on mycotoxin production and disease levels varied. Elevated CO_2 reduced DON levels in wheat grains but increased Fusarium crown rot incidence [61], while in maize it reduced fumonisin levels but enhanced susceptibility to *F. verticillioides* [62].

Disease resistance can interact with elevated CO_2 and other environmental conditions, as shown in Fusarium diseases of wheat (Tab. 1.1). In a FACE experiment, Melloy et al. [63] found that at elevated CO_2, crown rot interacted with moisture conditions and disease resistance with greater effect on the partially resistant cultivar. In the 2007 dry season, stem browning and pathogen biomass increased at elevated CO_2 in both susceptible cv. Tamaroi and partially resistant cv. 249 because *F. pseudograminearum* grows best under dry conditions of −1 mpa [64]. However in the 2008 wet season, pathogen biomass decreased at elevated CO_2 on both cvs. Tamaroi and 249 but the differences were not significant. In another FACE experiment over five wheat cropping cycles, Fusarium crown rot incidence increased with cropping cycles in susceptible cv. Tamaroi but not in partially resistant cv. 249 [61].

Induced resistance was observed in Fusarium crown rot in a FACE experiment of continuous wheat cropping, in which elevated CO_2 induced partial resistance in susceptible cv. Tamaroi but not in partially resistant cv. 249 (Tab. 1.1). However, the induced resistance was observed only in the first three of five cropping cycles, indicating that it is transient and inadequate to reduce crown rot or impede the selection and enrichment of highly aggressive strains in the pathogen population [61]. Induced transient resistance was also observed in another necrotrophic pathogen, *Colletotrichum gloeosporioides*, causing anthracnose disease of *Stylosanthes*, where resistance was induced in plants at elevated CO_2 but failed to operate when plants raised in a controlled environment at elevated CO_2 were exposed to pathogen inoculum under ambient CO_2 in the field [65]. The relationship between resistance and aggressiveness needs to be investigated, as aggressiveness interacted with resistance to *F. pseudograminearum* [61].

1.2.1.2 Elevated temperature

Temperature governs the growth and development rates of the different stages of the pathogen life cycle [66]. Elevated temperature can modify host physiology and resistance [44]. The effects of temperature on mycotoxigenic fungal growth and mycotoxin production *in vitro* and *in vivo* have been determined in numerous studies and summarized in many reviews [56, 67–69].

Projected higher temperatures in the next decades will possibly affect the prevalence of different mycotoxigenic fungal species and their growth, epidemiology, and mycotoxin production. *F. graminearum* favors the warm climate of southern Europe [52], and has become the dominant species responsible for Fusarium head blight in wheat in Europe, replacing *F. culmorum* based on surveys [16, 70–72]. This shift is possibly due to the higher optimum temperature range of *F. graminearum*. In Belgium, *F. graminearum* was predominant in 2004–2005 due to warm weather, while *F. culmorum* was dominant in 2001–2002 due to a low average July temperature of 17.4 °C [73]. For *F. verticillioides*, the risk of infection is high in areas with high temperatures (subtropical) than in temperate areas [74]. The systemic transmission of *F. verticillioides* from plant to kernels was increased at above average temperatures [75]. Fumonisin risk is typically higher at lower altitudes and latitudes due to warm conditions than high altitude and latitude regions [76].

Temperature influences the fungal growth and aflatoxin production of *Aspergillus flavus* in maize [56, 77]. The infection of *A. flavus* on maize ears is favored by high temperatures of 35–38 °C [78]. Paterson and Lima [77] stated that as temperature is projected to increase in temperate regions due to climate change, the risk of aflatoxin contamination increases based on the optimum temperature range of 30–33 °C. The percentage of highly toxigenic S strain of *A. flavus* communities increased as soil temperature increased from 16–32 °C, while the overall propagule density increased from 20–28 °C, suggesting that global warming would likely allow the S strain to dominate contaminated areas and increase aflatoxin contamination [79]. On the other hand, heat waves or extreme temperature can also be detrimental to aflatoxin production. A study showed that aflatoxin production was optimum at 28–30 °C but production stopped at 37 °C because aflatoxin biosynthesis was turned off [80].

1.2.1.3 Extreme rainfall

Above average rainfall has been attributed to the occurrence of Fusarium head blight epidemics. In North and South Dakota, USA in 1993, 250–600 % above normal precipitation was observed during a severe Fusarium head blight epidemic [81]. In the 2003 Fusarium head blight epidemic in southeastern USA, post-flowering rainfall was the most important weather variable that drove the epidemic [20]. Monthly average rainfall during anthesis and post anthesis was 10–150 % higher than long term averages when an outbreak of Fusarium head blight and DON contamination occurred in Australia in 2010 [82]. Similar severe epidemics of Fusarium head blight occurred in Norway from 2008–2012 when precipitation was above normal during anthesis and grain maturation [83].

Erratic rainfall patterns and high rain intensity can play a significant role in high levels of moisture in harvested grains leading to deterioration of grain quality, fungal growth, and mycotoxin contamination [55, 84, 85]. Unseasonable rains during harvest probably caused the aflatoxin contamination of maize stored under damp conditions which led to the 2004 Kenyan aflatoxicosis [28]. However, effects of climate change on stored grain are complex because moisture and temperature can interact with biotic factors such as insect pests to influence fungal species composition and dominance and mycotoxin production [55, 56, 77, 86].

1.2.1.4 Drought

The occurrence of drought episodes influences the growth and development of mycotoxigenic fungi and mycotoxin production in major food crops. In maize, environmental conditions leading to drought or water stress resulted in increased risk of fumonisin contamination by *F. verticillioides* but not *F. proliferatum* [87]. Severe wheat kernel rot due to *F. pseudograminearum* ensues when drought occurs post anthesis, as shown by “whiteheads”, shriveled grains, or no grains [88]. The occurrence of severe crown rot epidemics in drought-stressed environments may be due to an evolutionary response which recognizes polyamines in response to water stress. Polyamines are plant metabolites produced in response to abiotic stress such as water stress [89]. Spray inoculation of *F. graminearum* at mid-anthesis led to an increase in polyamines, fungal biomass, and DON in spike tissue [90].

Drought stress in addition to high temperature favors the growth, conidiation, and dispersal of *A. flavus* in maize with increased silk cut which compromises kernel integrity [91]. In Europe, hot and dry weather contributed to the 2003 outbreak of the previously uncommon *A. flavus* [55, 92]. Water stress also influences the composition of the communities of mycotoxin producers present. *A. flavus*, a more xerotolerant species than *F. verticillioides*, was able to colonize maize at the ripening stage by outcompeting the non-xerophilic *Fusarium* species [56]. *A. flavus* aflatoxin producers are associated with hot and dry agroecological zones [93].

1.2.1.5 Interactions of temperature and moisture

Quantitative knowledge of the interactions of climate change variables on mycotoxigenic fungi has been integrated into contour maps, for example the interaction of temperature and water availability on mycotoxigenic fungal growth and mycotoxin production [56, 69]. These data are important in generating baseline information for the understanding of basic relationships under climate change. For example, prediction of a 3 °C increase in temperature at water stressed conditions of 0.9 and 0.95 a_w showed no change in maximum growth rates of *F. verticillioides* and *F. graminearum* and a reduction in the maximum growth rate of *A. flavus* [56].

Temperature, moisture, and relative humidity are the most important climatic factors affecting Fusarium head blight development and DON accumulation [54, 94]. Fusarium head blight is favored by warm and wet weather at anthesis, resulting in partial or complete head blighting, reduced yield and quality, and mycotoxin production [84]. Window pane analysis using rank-based nonparametric correlation showed that the highest correlation was observed between Fusarium head blight intensity and 15–30 day windows (periods) during the final 60 day period (near anthesis) of combined moisture and temperature [95].

High temperature and drought stress are key factors for *A. flavus* growth and aflatoxin production [91, 96, 97]. Aflatoxin contamination occurs in two phases: during crop development due to physiologic stress or insect activity in the preharvest phase, and after maturation, either prior to harvest or postharvest during transport and storage in the second phase. Hot, dry conditions during crop development favor the first phase of contamination, while rain and warm temperatures with high humidity favor the second [98]. The developing maize crop is frequently resistant to *A. flavus* infection, but heat or drought stress and insect damage can compromise kernel integrity by increased “silk cut” [99]. Low rainfall and maximum temperature were found to be related to aflatoxin contamination in maize in various locations in Georgia, USA [100].

Relationships have been observed between teleconnections or variations in climate patterns in different regions of the globe and Fusarium head blight disease of wheat. The risk of Fusarium head blight was higher during El Niño or “neutral” years than La Niña years, when rainfall is higher in spring months in El Niño years [101]. In Ohio, USA, the Fusarium head blight epidemic risk occurs about a year following La Niña, while risk is low about a year after an El Niño episode. The epidemic can be location specific, however [95].

1.2.2 Pathogen population genetics and evolution

The impacts of climate change on the risk of mycotoxigenic fungal infection and mycotoxin production are complicated by concomitant effects of fungal species composition [76, 102] and evolutionary adaptation [48]. Shifts in the population composition of mycotoxigenic fungi have been observed, which may indicate adaptive evolution from nonrandom gene flow [48]. Fusarium head blight communities can produce different epidemics due to shifts in population composition which could be due to a changing climate [45, 103, 104]. Populations of *F. graminearum*, *F. pseudograminearum*, and *F. culmorum* are highly diverse in different countries and continents, and even individual fields with high gene flow, and therefore highly flexible in adapting to the environment as shown by rapid evolutionary changes on a large geographical scale [16]. The ability to evolve new phenotypes and genotypes via sexual recombination in *Aspergillus* might be accelerated by environmental stressors under climate change, such as intense heat and drought [105]. Most mycotoxins are mutagenic, and the rate of mycotoxin formation and mutation could increase under climate change [55].

Different environmental conditions may have driven epidemics of different *Fusarium* species. Canonical correlation analysis showed associations of *F. graminearum* with warm and humid conditions, *F. culmorum* with cooler, wet and humid conditions, and *F. poae* with drier and warmer conditions in wheat [106]. The increase in occurrence of *F. graminearum* in Europe may be due to its adaptation to cooler regions or climate change may have led to regions becoming warmer [102]. In the Netherlands in the 1980s–1990s, *F. culmorum* was the most predominant species but *F. graminearum* was more predominant in Western Europe from 2000 [16, 71, 103]. *F. asiaticum* occurs in warm regions > 15 °C while *F. graminearum* occurs in cooler regions in China [107].

Different environmental conditions affect composition of the *Fusarium* species in maize. In Europe, red ear rot (fusariosis), primarily caused by *F. graminearum*, results in contamination of ears with DON, zearalenone (ZEA), and nivalenol (NIV) [108]. It is severe in locations and years with frequent rainfall and low temperatures in summer and early fall [109]. Pink ear rot, caused by *F. verticillioides* and *F. proliferatum*, occurs frequently in warm, dry conditions, while pink ear rot, caused by *F. subglutinans*, occurs in cold, humid conditions [110]. Drought stress and temperatures above 25 °C favor *F. verticillioides* over *F. graminearum* in maize [94].

Shifts in the *Fusarium* chemotype or strain producing a type of trichothecene type B toxin were observed in recent epidemics which can be attributed to changes in environmental conditions. DON chemotype strains of *F. graminearum* are classified into two groups: (1) DON-chemotype 1A, producing DON and 3-acetylated deoxynivalenol (3ADON), mostly European strains from warmer regions, and (2) DON chemotype 1B producing DON and 15-acetylated deoxynivalenol (15ADON), mostly American strains from slightly cooler regions [111].

The rapid spread of a highly toxigenic *F. graminearum* population was observed in North America, in which 3ADON population chemotype frequency increased > 14 fold from 1998–2004. However, the basis for the shift remains unclear, although changes in agricultural practices or environmental conditions may have driven the shift in Fusarium head blight pathogen composition [112]. On exposure to extreme high and low temperatures, *F. graminearum* 3ADON showed faster mycelial growth and produced more DON and ZEA than 15ADON [113]. A similar shift was observed in China, in which the 3ADON producer of *F. asiaticum* increased significantly in the middle valley of the

Yangtze River between 1999 and 2005. This highly toxigenic and aggressive chemotype population is spreading east to west in China, which may be due to natural selection [114]. In controlled conditions, recovery of 15ADON was predominant at 18–22 °C and 3ADON at 28 °C, and mixed at 24 °C. However, no correlation with chemotype and temperature or precipitation was found [115].

The shifts from 15ADON to 3ADON chemotypes may be due to a general fitness advantage, as shown by the higher growth rate and spore production observed by Ward et al. [112]. However, no detectable advantage of 3ADON isolates over 15ADON isolates in saprophytic or pathogenic fitness was observed on a susceptible wheat cultivar [116]. Different components of host-parasite fitness such as infectivity and fecundity are differentially altered by the environment [51]. The shifts are unlikely to result in higher aggressiveness but indicated by higher DON levels in moderately resistant or susceptible genotypes [117]. Yet one study showed that the 3ADON chemotype was more aggressive than the 15ADON and NIV chemotypes [118], the latter being mostly associated with maize [119]. In an earlier study using a crossing population of *F. graminearum*, DON producing progeny were twice as aggressive as NIV producers [120]. Different inoculation methods and/or environmental conditions may have caused the differences in the recent work [118], hence more research is needed.

Aggressiveness as a quantitative trait is highly affected by temperature, rainfall, and relative humidity; and variation in aggressiveness among and within populations partly determines adaptation and fitness of the pathogen [121, 122]. For example, in a cross between two European DON-producing isolates of *F. graminearum*, the progeny × environmental interaction was one of the most important sources of variance, accounting for 29 % of total variance for aggressiveness and 19 % for DON production, suggesting the importance of multi-environmental trials [121].

The strains or morphotypes of *A. flavus* are adapted to distinct ecological niches. *A. flavus* S strain produce more aflatoxin than the L strain isolates [79], and was the primary cause of the aflatoxin contamination in the 2004 Kenyan epidemic [123]. In the Sonoran desert, Arizona, USA, *A. flavus* S strains dominated during the warmest period and L strain during winter and spring [124]. The incidence of *A. flavus* S strain was also higher in warm than in cold seasons in South Texas, USA [79]. *A. flavus* S_{BG} strain (unknown aflatoxin B and G producing strain) was more prevalent in areas with the highest average temperature [93] and had the highest frequency in crops from semi-arid and sub-humid parts of West Africa [125]. The variation of S_{BG} incidence across agroecological zones may be caused partly by crop rotation [126].

1.3 Prediction of climate change effects on epidemics

Forecasting systems are needed for strategic long term decisions [127] such as addressing the effects of climate change on plant diseases [45]. Predictive models will provide the foresight for strategic climate change adaptation and policy guidance [52]. In order to use plant disease models in forecasting climate change effects, a quantitative understanding of plant disease epidemiology and rigorous model verification and validation are warranted [128]. Scientifically valid forecasts of climate change effects on plant diseases are needed to formulate robust food security policies [129].

Quantitative approaches dealing with scenarios and interactions are needed for impact assessment of climate change effects on plant health [44, 45]. Realistic climate change scenarios are needed in predictive modeling of risks of mycotoxigenic fungal growth and mycotoxin production [77]. However, typical scenario analysis may be limited by simplistic assumptions and there is a need to use more complete scenarios incorporating thresholds, interactions, and feedback loops [45]. *In vitro* studies on interacting factors of elevated CO_2, temperature, and/or moisture on mycotoxigenic fungal growth and mycotoxin production have been investigated [56]; this can provide baseline information for simulation modeling [55].

1.3.1 Bioclimatic niche models

Ecological niche or bioclimatic envelope models are correlative models which predict species distributions with climate variables or through an understanding of species' physiological responses to climate change [130]. The bioclimatic envelope approach can provide a useful first approximation of the potential impact of climate change [131]. This modeling technique can be used to investigate the effect of climate change on plant diseases based on climatic data [9], but caution is needed in the assumption of the climate limits of species distribution and representation of uncertainties [132].

The effect of climate change on mycotoxin risk was projected in major maize growing provinces in the Philippines [133] as a case study (Fig. 1.1), and related to food production and security. Bioclimatic envelope modeling was used to predict the preharvest risks of Aspergillus and Fusarium maize ear rots during dry (first) and wet (second) cropping seasons based on the projected temperature increase for the year 2050 under the RCP8.5 climate change scenario (Fig. 1.2). The model was developed based on temperature ranges of *A. flavus* and *F. verticillioides* from Sumner and Lee [134] and Stewart et al. [135], respectively, using R programming [136]. Aflatoxin contamination in agricultural crops is a serious problem in the Philippines because of high temperatures and high relative humidity [137, 138], while *F. verticillioides* is the predominant fumonisin-producing *Fusarium* species in the country [139, 140].

Comparison of current and predicted distributions of Aspergillus ear rot showed an increased risk in 2050 in the Philippines (Fig. 1.3). As expected, higher risk was estimated during dry season compared to wet season cropping. Under climate change, medium risk was projected for the province of North Cotabato and Maguindanao, while high risk was projected for the province of Isabela during dry season cropping

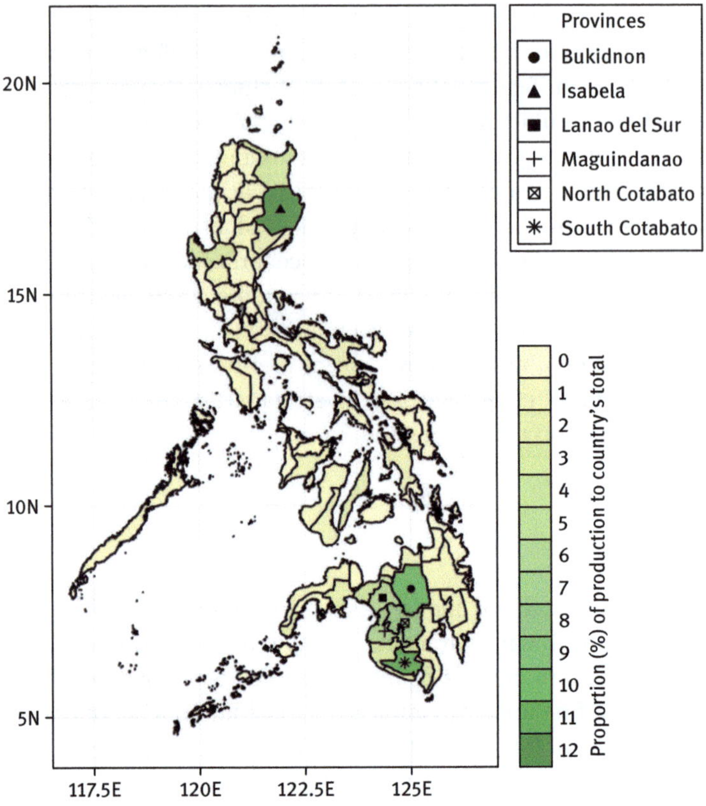

Fig. 1.1: Map of the Philippines showing major maize producing provinces and their contribution (%) to the country's total maize production.

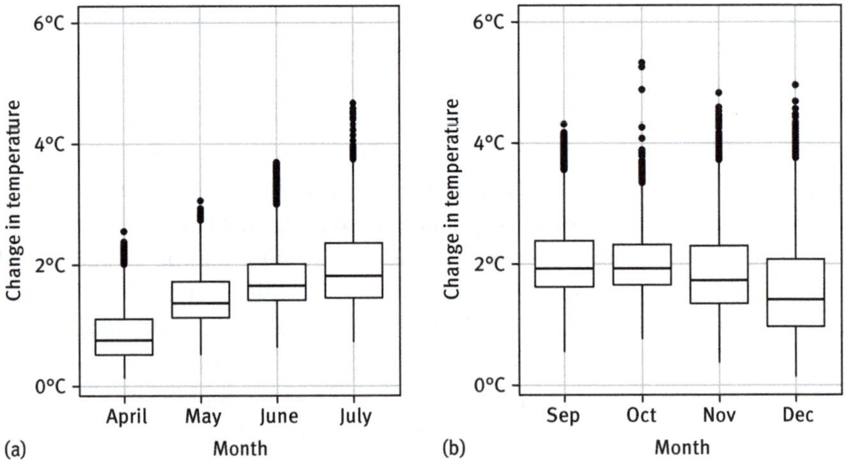

Fig. 1.2: Projected change in average monthly temperature for the 1st (a) and 2nd (b) cropping seasons of maize in the Philippines based on IPCC AR5 climate scenario RCP8.5.

(Tab. 1.2). These results can be explained by the high temperature requirement for Aspergillus ear rot development observed in the dry season [78].

Comparison of current and predicted distributions of Fusarium ear rot showed increased risks in 2050 (Fig. 1.4). The current Fusarium risks are medium to high and low to medium in the dry and wet seasons, respectively. There will be medium to high risk of Fusarium ear rot infection in both dry and wet seasons. In the major maize growing areas in the dry season, there will be high risk in Isabela, North Cotabato and Maguindanao, and medium risk in South Cotabato. However, in the wet season there will be high risk in North Cotabato and Maguindanao, and medium risk in Isabela and South Cotabato (Tab. 1.2). These results showed increased Fusarium ear rot risks in both dry and wet seasons because high temperatures are favorable for Fusarium ear rot development [135].

These results have significant implications regarding the risk of mycotoxigenic fungi on maize which could affect food security under climate change. Fumonisin production of *F. verticillioides* isolates was significantly higher in Isabela province as compared to Laguna province [140]. Maguindanao province, with a projected high risk for Fusarium ear rot and medium risk for Aspergillus ear rot, has the highest proportion of families (84.5 %) spending more than 50 % of their total expenditure on food. Isabela province, with high risk for Aspergillus ear rot and Fusarium ear rot, has 64.5 % of families spending more than 50 % of their total expenditure on food [141].

Tab. 1.2: Projected Aspergillus and Fusarium ear rot preharvest risks for major maize producing areas in the Philippines based on projected temperature increase for the year 2050 under IPCC climate change scenario RCP8.5.

Top maize producing provinces		Isabela	South Cotabato	Bukidnon	North Cotabato	Maguin-danao	Lanao del Sur
Proportion of country's total production (%)		12	11	10	7	6	5
Average annual production (metric tons)[a]		641 936	583 631	547 155	396 812	335 241	297 977
Aspergillus ear rot	**1st Cropping**	High	Very low	Very low	Medium	Medium	Very low
	2nd Cropping	Low	Very low	Very low	Low	Low	Very low
Fusarium ear rot	**1st Cropping**	High	Medium	Low	High	High	Very low
	2nd Cropping	Medium	Medium	Low	High	High	Very low

a Data source: Philippine Statistical Authority-Bureau of Agricultural Statistics [133]

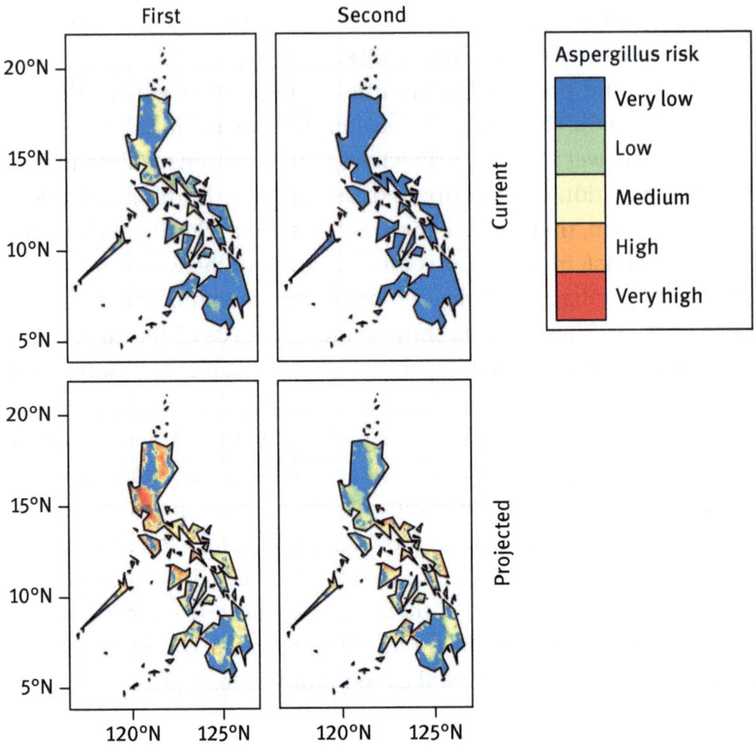

Fig. 1.3: Current and projected preharvest risk of Aspergillus ear rot development during the first and second cropping seasons in the Philippines in 2050 based on projected temperature increase for the year 2050 under the IPCC AR5 climate scenario RCP8.5. Risk categories are: very low (1–20 %), low (21–40 %), medium (41–60 %), high (61–80 %), and very high (81–100 %).

1.3.2 Climate change scenario models

The modeling approach to simulating plant disease development under future climate change needs to address model uncertainty and complexity. Uncertainty in model inputs can compromise the reliability of simulation results due to the propagation of errors [44]. To reduce uncertainty, multiple models [45] or a comparison of different models can be implemented [142]. Climate change effects on plant health can have direct effects of climate variables and indirect effects via global changes such as cropping practices [9, 47], leading to system complexity [46]. An analytical framework was developed to evaluate and improve the plant disease model to address complexity. For Fusarium head blight, complexity arises from having multiple host species, the importance of spatial components, risk from crop residues, and potential to reduce ecosystem services such as the production of food and efficiency of conversion of natural resources into food [45].

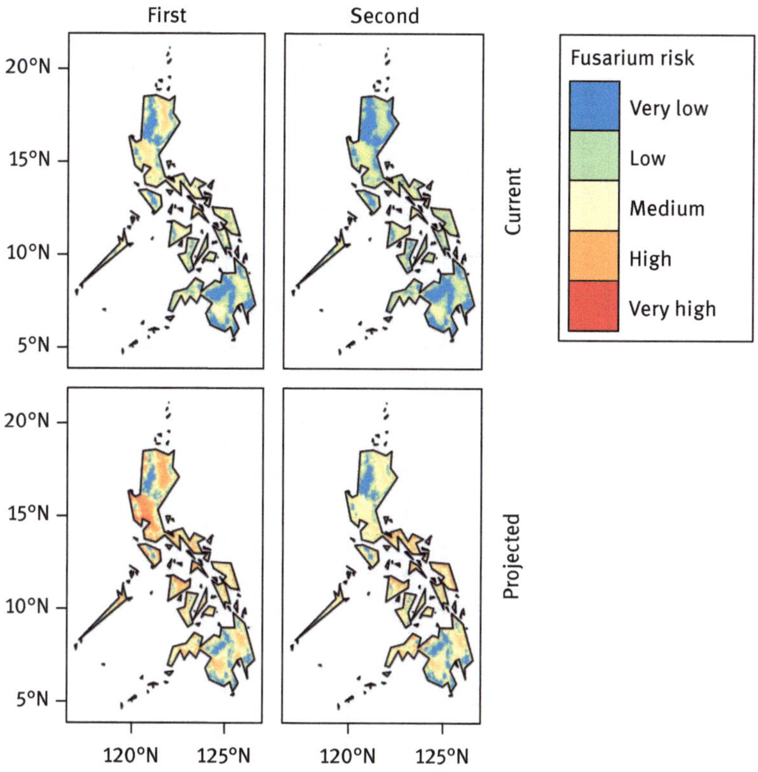

Fig. 1.4: Current and projected preharvest risk of Fusarium ear rot development during the first and second cropping seasons in the Philippines in 2050 based on projected temperature increase for the year 2050 under the IPCC AR5 climate scenario RCP8.5. Risk categories are: very low (1–20 %), low (21–40 %), medium (41–60 %), high (61–80 %) and very high (81–100 %).

Predictions of risks of disease intensity and mycotoxin contamination under climate change from different studies are shown in Table 1.3. Qualitative assessments based on expert knowledge of epidemiology indicated no increase [143–145], but this method serves as an initial step in climate change impact assessment. The simulation results using coupled climate disease and crop models showed increased [146–148] or decreased risks [149]. Simulated risk variability [150, 151] and inconsistency among studies indicate the influences of model uncertainties and spatial and climatic variation between regions [47, 50, 152]. Application of current Fusarium head blight disease and mycotoxin models in new regions or countries (models developed elsewhere) yielded poor results, indicating site and year specificity as in descriptive models [82, 146, 148, 153] prompting the need for thorough field validation of models before use. The comparison of simulation results is difficult across different approaches in climate change research considering spatial and temporal variations in many interacting factors such as crop development, cultivar, isolate or species, and climate [152].

Tab. 1.3: A summary of mycotoxigenic fungal disease and mycotoxin predictive models used to predict climate change impacts.

Country	Disease	Crop	Model	Prediction	Reference
Brazil	Fusarium head blight (*Fusarium graminearum*)	Wheat	Linked process based model	Increase in Passo Fundo and Estanzuela; no change in Pergamino	Fernandes et al. [150]
Canada	Fusarium head blight (*Fusarium* spp.)	Wheat	Qualitative assessment	No change	Boland et al. [143]
Canada	Fusarium ear rot	Maize	Qualitative assessment	Increase	Boland et al. [143]
UK	Fusarium head blight	Wheat	Qualitative assessment (Ecotype)	Slight increase	Madgwick et al. [144]
Sweden	Fusarium head blight	Wheat	Qualitative assessment	Increase in mycotoxin levels due to more humid climate	Roos et al. [145]
UK	Fusarium head blight	Wheat	Combined crop and disease model	Increase in Southern England by 2050s	West et al. [168]
Europe	*Aspergillus flavus* Aflatoxin	Maize Wheat Rice	*A .flavus* – AFB1 model	Risk increase in maize Low risk in wheat Absent in rice	Battilani et al. [151]
Europe	Fusarium head blight DON	Wheat	DON prediction model	Increase in DON levels by a factor of 3	Van der Fels-Klerx et al. [147]
Netherlands	Fusarium head blight DON	Winter wheat Maize	DON prediction model	Decrease (HG model) and no effect (KNM1 climate model) in maize Slight decrease in DON in wheat (both climate models)	Van der Fels-Klerx et al. [149]
China	Fusarium head blight	Wheat	Logistic weather based regression model	Increase in southern locations in Anhui province in 2020–2050	Zhang et al. [148]

1.4 Management of plant diseases caused by mycotoxigenic fungi under climate change

When ensuring food security under climate change, the primary basis of developing effective plant disease management strategies in major food crops is realistic prediction of climate change impacts on plant diseases. The combined effects of climate and global changes may have unexpected consequences [9]. Modeling efforts should be able to reduce model uncertainty and address complexity [45]. Mechanistic incorporation of pathogens in crop models could produce more realistic predictions of crop production on a regional scale which could aid the formulation of appropriate food security policies [34, 129].

Investment in adaptation and mitigation strategies is necessary to prevent the negative impacts of climate change on plant diseases via crop losses hampering progress in achieving food security. Resilience to climate change requires an integrated approach which addresses all dimensions of the food system [32], with particular attention to reducing climate change vulnerability [154]. A practical approach is to prepare for all likely eventualities, but risks based on probabilities generated by stochastic models can better guide a judicious and effective development of management options [155]. Thus, diverse disease management strategies can be used involving participatory and interdisciplinary approaches [47]. There is a need to evaluate the efficacy of current disease management strategies and develop new disease management tools and methods [84].

Reducing the level of infection of mycotoxigenic fungi and associated mycotoxin production in grains is the foremost requirement in securing crop yields to guarantee food security. Necrotrophic pathogens such as *Aspergillus* and *Fusarium* species have wide host ranges and do not follow gene-for-gene specificity, thus lacking major genes for resistance [27, 156]. Fusarium head blight resistance is a complex quantitative trait which depends on environmental conditions. Partial resistance needs to be combined with agronomic practices to develop robust integrated disease management strategies [84, 156]. Breeding for passive Fusarium head blight resistance based on morphological traits such as canopy and spike architecture involving plant height and anther extrusion, and physiological traits such as flowering date [156] need to consider the modifying effects of elevated CO_2 and temperature on disease resistance, canopy architecture, and microclimate [157, 158]. When the environment is highly favorable for Fusarium head blight infection, implementation of a single management strategy fails and the use of disease resistance and fungicides is necessary to achieve the 75 % reduction in Fusarium head blight index required to manage the disease [159].

1.5 Outlook and conclusions

An improved understanding of the effects of climate change on biological and food systems is a critical step towards addressing the impacts of climate change on food security [160], which includes the effects on pests, pathogens, and weeds. The changing geographic distribution of mycotoxigenic fungal species composition and chemotypes in cereals and related accelerated evolution under climate change are paramount research topics which need to be addressed [48]. Research is also needed to study probable mycotoxigenic fungal shifts in rice due to the recent emergence of *Fusarium fujikuroi* causing bakanae disease, previously a minor disease, in the Philippines [161, 162], and infection of rice by *Fusarium proliferatum* in Costa Rica [163]. In the first study, the most important sources of variation of aggressiveness of *F. fujikuroi* are isolate and isolate-environment interaction accounting for about 34 % and 42 %, respectively, of the total variance for PSBRc82 and IR42 varieties [162]. Basic research is needed to study the effects of the main climate change variables: elevated CO_2, temperature, and moisture and their interactions on mycotoxigenic fungal diseases and mycotoxins since comprehensive knowledge is still lacking [55, 56, 76].

Realistic predictions of climate change impact on mycotoxin and disease risk through the development of reliable predictive models are imperative to minimize production losses and achieve food security. Approaches to model improvement should address model uncertainty and complexity [46, 47, 50]. Monitoring and 'early warning' of mycotoxins and application of geographic information systems and geostatistics are important research needs in surveillance [9, 56, 164, 165].

The formulation of adaptive disease management strategies is the key to ensuring food security under climate change. A shift to adaptation and mitigation strategies is needed in addition to impact assessment. Current disease management strategies should be evaluated for climate change adaptation but novel technologies will be very helpful [84, 166, 167].

References

[1] IPCC. Summary for policymakers. In: Stocker TF, Qin D, Plattner GK, et al., eds. *Climate Change 2013: The Physical Science Basis*. Contribution of Working Group I to the Fifth Assessment Report of the Intergovernmental Panel on Climate Change Cambridge, United Kingdom and New York, NY, USA: Cambridge University Press; 2013.

[2] IPCC. Summary for policymakers. In: Field CB, Barros VR, Dokken DJ, et al., eds. *Climate Change 2014: Impacts, Adaptation, and Vulnerability. Part A: Global and Sectoral Aspects.* Contribution of Working Group II to the Fifth Assessment Report of the Intergovernmental Panel on Climate Change Cambridge, United Kingdom and New York, NY, USA: Cambridge University Press; 2014. p. 1–32.

[3] Godfray HCJ, Beddington JR, Crute IR, et al. Food security: the challenge of feeding 9 billion people. *Science* 2010;327:812–818.

[4] Lal R. Food security in a changing climate. *Ecohydrology and Hydrobiology* 2013;13:8–21.

[5] Porter M, Xie L, Challinor AJ, et al. Food Security and Food Production Systems. In: Field CB, Barros VR, Dokken DJ, et al., eds. *Climate Change 2014: Impacts, Adaptation, and Vulnerability. Part A: Global and Sectoral Aspects.* Contribution of Working Group II to the Fifth Assessment Report of the Intergovernmental Panel on Climate Change Cambridge, United Kingdom and New York, NY, USA: Cambridge University Press; 2014. p. 1–82.

[6] Rosenzweig C, Iglesias A, Yang XB, Epstein PR, Chivian E. Climate change and extreme weather events: Implications for food production, plant diseases, and pests. *Global Change and Human Health* 2001;2:90–104.

[7] Rosenzweig C. Climate change and agriculture. In: Meyers RA, ed. *Extreme Environmental Events: Complexity in Forecasting and Early Warning*. Springer Reference 2011. p. 31–41.

[8] Oerke EC. Crop losses to pests. *J Agr Sci* 2006; 144: 31–43.

[9] Savary S, Nelson A, Sparks AH, et al. International agriculture research tackling the effects of global and climate changes on plant diseases in the developing world. *Plant Dis* 2011;95:1204–1216.

[10] Hellin J, Shiferaw B, Cairns JE, et al. Climate change and food security in the developing world: Potential of maize and wheat research to expand options for adaptation and mitigation. *Journal of Development and Agricultural Economics* 2012;4:311–321.

[11] Esker PD, Savary S, McRoberts N. Crop loss analysis and global food supply: focusing now on required harvests. *CAB Reviews: Perspectives in Agriculture, Veterinary Science, Nutrition and Natural Resources* 2012;7:1–14.

[12] Savary S, Ficke A, Aubertot JN, Hollier C. Crop losses due to diseases and their implications for global food production losses and food security. *Food Sec* 2012;4:519–537.

[13] Bottalico A, Perrone G. Toxigenic *Fusarium* species and mycotoxins associated with head blight in small-grain cereals in Europe. *Eur J Plant Pathol* 2002; 108: 611–624.

[14] Osborne LE, Stein JM. Epidemiology of Fusarium head blight on small-grain cereals. *Int J Food Microbiol* 2007;119:103–108.

[15] Chakraborty S, Liu CJ, Mitter V, et al. Pathogen population structure and epidemiology are keys to wheat crown rot and Fusarium head blight management. *Australas Plant Path* 2006;35:643–455.

[16] Miedaner T, Cumagun CJR, Chakraborty S. Population genetics of three important head blight pathogens *Fusarium graminearum*, *F. pseudograminearum* and *F. culmorum*. *J Phytopathol* 2008;156:129–139.

[17] McMullen M, Jones R, Gallenberg D. Scab of wheat and barley: A re-emerging disease of devastating impact. *Plant Dis* 1997;81:1340–1348.

[18] McMullen M, Bergstrom G, De Wolf ED, et al. A unified effort to fight an enemy of wheat and barley: Fusarium head blight. *Plant Dis* 2012;96:1712–1728.

[19] Nganye WE, S. K, Wilson WW, Leistritz FL, Bangsind DA. Economic Impacts of Fusarium Head Blight in Wheat and Barley: 1993–2001. Fargo, Narth Dakota: Agricultural Experiment Station North Dakota State University, 2004.

[20] Cowger C, Sutton AL. The southwestern U.S. Fusarium head blight epidemic of 2003. *Plant Health Prog* 2005.

[21] Charmley LL, Trenholm HL, Prelusky DA, Rosenberg A. Economic losses and decontamination. *Nat Toxins* 1995; 3: 99–203.

[22] Bryden WI. Mycotoxins in the food chain: human health implications. *Asia Pac J Clin Nutr* 2007;16:95–101.

[23] CAST. Mycotoxins: Risks in plant, animal, and human systems systems. Ames, Iowa, Council for Agricultural Science and Technology, 2003.

[24] Wu F. Mycotoxin reduction in Bt corn: potential economic, health, and regulatory impacts. *Transgenic Res* 2006;15:277–289.
[25] Institute M. Economic impact of aflatoxin: Aflatoxin's negative impact on trade. 2014.
[26] Wise K. Diseases of Corn: Corn Disease Loss Estimates From the United States and Ontario, Canada - 2012: Purdue University, 2013.
[27] Marasas WF, Gelderbrom WCA, Shephard GS, Vismer HF. Mycotoxins: A global problem. In: Leslie JF, Bandyopadhyay R, Visconti A, eds. *Mycotoxins: Detection Methods, Management, Public Health and Agricultural Trade*. Wallingford, UK: CAB International; 2008. p. 29–39.
[28] Lewis L, Onsongo M, Njapau H, Schurz-Rogers H, Luber G, Kieszak S. Aflatoxin contamination of commercial maize products during an outbreak of acute aflatoxicosis in Eastern and Central Kenya. *Environ Health Persp* 2005;113:1763–1777.
[29] Lubulwa ASG, Davis JS. Estimating the social costs of the impacts of fungi and aflatoxins in maize and peanuts. In: Highley E, Wright EJ, Banks HJ, Champ BR, eds. *Stored product protection: Proceedings of the 6th International Working Conference on Stored-product Protection*, Wallingford, UK: CAB International: 1994.
[30] FAO, WFP, IFAD. The State of Food Insecurity in the World 2012. Economic Growth is Necessary but Not Sufficient to Accelerate Reduction of Hunger and Malnutrition. Rome, Italy: FAO; 2012.
[31] Lobell DB, Burke MB, Tebaldi C, Mastrandea MD, Falcon WP, Naylor RL. Prioritizing climate change adaptation needs for food security in 2030. *Science* 2008;319:607–610.
[32] Wheeler T, von Braun J. Climate change impacts on global food sesurity. *Science* 2013;341:508–513.
[33] Vermeulen SJ, Campbell BM, Ingram JSI. Climate change and food systems. *Ann Rev Environ Resour* 2012;37:195–222.
[34] Gregory PJ, Johnson RH, Newton AC, Ingram JSI. Integrating pests and pathogens into the climate change/food security debate. *J Exp Bot* 2009;60:2827–2838.
[35] Nelson GC, Rosegrant MW, Palazzo A, et al. Food Security, Farming, and Climate Change to 2050: Scenarios, Results, Policy Options. Washington, DC: International Food Policy Research Institute; 2010.
[36] FAO. Rome Declaration and World Food Summit Plan of Action. Rome, Italy: FAO; 1996.
[37] Flood J. The importance of plant health to food security. *Food Sec* 2010; 2: 215–231.
[38] Carefoot GL, Sprott ER. Famine on the Wind Chicago: Rand McNally; 1967.
[39] Padmanabhan SY. The great Bengal famine. *Ann Rev Phytopathol* 1973;11:11–26.
[40] Singh RP, Hodson DP, Huerta-Espino J, et al. The emergence of Ug99 races of the stem rust fungus is a threat to world wheat production. *Ann Rev Phytopathol* 2011;49:465–481.
[41] Strange RN, Scott PR. Plant disease: A threat to global food security. *Ann Rev Phytopathol* 2005;43:83–116.
[42] Pinstrup-Andersen P. Food security: definition and measurement. *Food Sec* 2000;1:5–7.
[43] Zeigler RS, Savary S. Plant diseases and the world's dependence on rice. In: Strange RN, Gullino ML, eds. *The Role of Plant Pathology in Food Safety and Food Security*. Dordrecht: Springer; 2010. p. 3–9.
[44] Coakley SM, Scherm H, Chakraborty S. Climate change and plant disease management. *Ann Rev Phytopathol* 1999;37:399–426.
[45] Garrett KA, Forbes GA, Savary S, et al. Complexity in climate-change impacts: an analytical framework for effects mediated by plant disease *Plant Pathol* 2011;60:15–30.
[46] Juroszek P, Tiedemann AV. Plant pathogens, insect pests, andweeds in a changing global climate: a review of approaches, challenges, research gaps, key studies and concepts. *J Agr Sci* 2012.
[47] Pautasso M, Doring TF, Garbelotto M, Pellis L, Jeger MJ. Impacts of climate change on plant diseases – opinions and trends. *Eur J Plant Pathol* 2012; 133: 295–313.

[48] Chakraborty S. Migrate or evolve: options for plant pathogens underclimate change. *Glob Change Biol* 2013;19:1985–2000.

[49] Jump AS, Penuelas J. Running to stand still: Adaptation and the resopnse of plants to rapid climate change. *Ecol Lett* 2005;8:1010–1020.

[50] Garrett KA, Dendy SP, Frank EE, Rouse MN, Travers SE. Climate change effects on plant disease: Genomes to ecosystems. *Ann Rev Phytopathol* 2006; 44: 489–509.

[51] Wolinksa J, King KC. Environment can alter selection in host-parasite interactions. *Trends Parasitol* 2009;25:236–244.

[52] Miraglia M, Marvin HJP, Kleter GA, et al. Climate change and food safety: An emerging issue with special focus on Europe. *Food Chem Toxicol* 2009;47:1009–1021.

[53] Magan N. Fungi in extreme environments. In: Kubicek CP, Drizhinina IS, eds. *Environmental and Microbial Relationships. The MYCOTA IV.* 2nd ed. Berlin, Germany: Springer Verlag; 2007. p. 85–103.

[54] Wegulo SN. Factors influencing deoxynivalenol accumulation in small grain cereals. *Toxins* 2012;4:1157–1180.

[55] Paterson RRM, Lima N. Further mycotoxin effects from climate change. *Food Res Int* 2011;44:2555–2566.

[56] Magan N MA, Aldred D. Possible climate-change effects on mycotoxin contamination of food crops pre- and postharvest. *Plant Pathol* 2011;60:150–163.

[57] Samapundo S, De Meulenaer B, Atukwase A, Debevere J, Devlieghere F. The influence of modified atmospheres and their interaction with water activity on the radial growth and fumonisin B1 production of *Fusarium verticillioides* and *F. proliferatum* on corn. Part I: The effect of initial headspace carbon dioxide concentration. *Int J Food Microbiol* 2007;114:160–167.

[58] Hocking AD. Responses of fungi to modified atmospheres. Fumigation and Controlled Atmosphere Storage of Grain. *Proceedings of an International Conference* 1989; Singapore. p. 70–82.

[59] Chakraborty S, Luck J, Hollaway G, et al. Impacts of global change on diseases of agricultural crops and forest trees. *CAB Reviews: Perspectives in Agriculture, Veterinary Science, Nutrition and Natural Resources* 2008;3:1–15.

[60] Eastburn DM, McElrone AJ, Bilgin DD. Influence of atmospheric and climatic change on plant-pathogen interactions. *Plant Pathol* 2011;60:54–69.

[61] Khudhair M, Melloy P, Lorenz DJ, et al. Fusarium crown rot under continuous cropping of susceptible and partially resistant wheat in microcosms at elevated CO_2. *Plant Pathol* 2013;1–11.

[62] Vaughan MM, Huffaker A, Schmelz EA, et al. Effects of elevated [CO_2] on maize defence against mycotoxigenic *Fusarium verticillioides*. *Plant Cell Environ* 2014.

[63] Melloy P, Hollaway G, Luck J, Norton R, Aitken E, Chakraborty S. Production and fitness of *Fusarium pseudograminearum* inoculum at elevated carbon dioxide in FACE. *Glob Change Biol* 2010;16:3363–3373.

[64] Singh DP, Backhouse D, Kristiansen P. Interactions of temperature and water potential indisplacement of *Fusarium pseudograminearum* from cereal residues by fungal antagonists. *Biol Control* 2009;48:188–195.

[65] Pangga IB, Chakraborty S, Yates D. Canopy size and induced resistance in *Stylosanthes scabra* determine anthracnose severity at high CO_2. *Phytopathology* 2004;94:221–227.

[66] Agrios GN. *Plant Pathology* 5th ed. UK: Elsevier; 2005.

[67] Doohan FM, Brennan J, Cooke BM. Influence of climatic factors on *Fusarium* species pathogenic to cereals. *Eur J Plant Pathol* 2003;109:755–768.

[68] Champeil A, Doré T, Fourbet JF. Fusarium head blight: epidemiological origin of the effects of cultural practices on head blight attacks and the production of mycotoxins by *Fusarium* in wheat grains. *Plant Sci* 2004;166:1389–1415.

[69] Sanchis V, Magan N. Environmental conditions affecting mycotoxins. In: Magan N, Olsen M, eds. Mycotoxins in Food: Detection and Control. Boca Raton: CRC Press; 2004. p. 174–179.

[70] Waalwijk C, Kastelein P, Vries PM, et al. Major changes in *Fusarium* spp. in wheat in the Netherlands. *Eur J Plant Pathol* 2003;109:743–754.

[71] Jennings P, Coates ME, Turner JA, Chandler EA, Nicholson P. Determination of deoxynivalenol and nivalenol chemotypes of *Fusarium culmorum* isolates from England and Wales by PCR assay. *Plant Pathol* 2004;53:182–190.

[72] Yli-Mattila T, Rämö S, Hietaniemi V, Hussien T, Lopez ALC, Cumagun CJR. Molecular quantification and genetic diversity of toxigenic *Fusarium* species in Northern Europe as compared to those in southern Europe. *Microorganisms* 2013;1:162–174.

[73] Isebaert S, De Saeger S, Devreese R, Verhoeven R, Maene P, Heremans B. Mycotoxin-producing *Fusarium* species occurring in winter wheat in Belgium (Flanders) during 2002–2005. *J Phytopathol* 2009;157:108–16.

[74] de la Campa R, Hooker DC, Miller JD, Schaafsma AW, Hammond BC. Modelling effects of environment, insect damage, and Bt genotypes on fumonisin accumulation in maize in Argentina and the Philippines. *Mycopathologia* 2005;154:539–552.

[75] Murillo-Williams A, Munkvold G. Systemic infection by *Fusarium verticillioides* in maize plants grown under three temperature regimes. *Plant Dis* 2008;92:1695–1700.

[76] Wu F, Bhatnagar D, Bui-Klime T, et al. Climate change impacts on mycotoxin risks in US maize. *World Myco J* 2011;4:79–93.

[77] Paterson RRM, Lima N. How will climate change affect myctotoxins in food? *Food Res Int* 2010;43:1902–1914.

[78] Jones R, Duncan H, Payne G, Leonard K. Factors influencing infection by *Aspergillus flavus* in silk-inoculated corn. *Plant Dis* 1980;64:859–863.

[79] Jaime-Garcia R, Cotty PJ. Crop rotation and soil temperature influence the community structure of *Aspergillus flavus* in soil. *Soil Biol Biochem* 2010;42:1842–1847.

[80] OBrian GR, Georgianna DR, Wilkinson JR, et al. The effect of elevated temperature on gene transcription and aflatoxin biosynthesis. *Mycologia* 2007;99:232–239.

[81] Wiersma JV, Peters EL, Hanson MA, Bouvette RJ, Busch RH. Fusarium head blight in hard red spring wheat: cultivar responses to natural epidemics. *Agron J* 1996;88:223–230.

[82] Obanor F, Neate S, Simpfendorfer S, Sabburg R, Wilson P, Chakraborty S. *Fusarium graminearum* and *Fusarium pseudograminearum* caused the 2010 headblight epidemics in Australia. *Plant Pathol* 2013;62:79–91.

[83] Sundheim L, Brodal L, Hofgaard IS, Rafoss T. Temporal variation of mycotoxin producing fungi in Norwegian cereals. *Microorganisms* 2013;1:188–198.

[84] Chakraborty S, Newton AC. Climate change, plant diseases and food security: an overview. *Plant Pathol* 2011;60:2–14.

[85] Marroquin-Cardona AG, Johnson NM, Philips TD, Hayes AW. Mycotoxins in a changing environment - A review. *Food Chem Toxicol* 2014;69:220–230.

[86] Magan M, Hope R, Cairns V, Aldred D. Post-harvest fungal ecology: Impact of fungal growth and mycotoxin accumulation in stored grain. *Eur J Plant Pathol* 2003;109:723–730.

[87] Marin P, Magan N, Vazquez C, Gonzalez-Jaen MT. Differential effect of environmental conditions on the growth and regulation of the fumonisin biosynthetic gene *FUM1* in the maize pathogens and fumonisin producers *Fusarium verticillioides* and *Fusarium proliferatum*. *FEMS Microbial Ecol* 2010;73:303–311.

[88] Luck J, Spackman M, Freeman A, et al. Climate change and diseases of food crops. *Plant Pathol* 2011;60:113–121.

[89] Tunali B, Obanor F, Erginbas G, Westecott RA, Nicol J, Chakraborty S. Fitness of three *Fusarium* pathogens of wheat. *FEMS Microbial Ecol* 2012;81:596–609.

[90] Gardiner D, Kazan K, Praud S, Torney F, Rusu A, Manners J. Early activation of wheat polyamine biosynthesis during Fusarium head blight implicates putrescine as an inducer of trichothecene mycotoxin production. *BMC Plant Biol* 2010;10:289.

[91] Cotty J, Jaime-Garcia R. Influence of climate on aflatoxin producting fungi and aflatoxin contamination. *Int J Food Microbiol* 2007;119:109–115.

[92] Giorni P, Magan N, Pietri A, Bertuzzi T, Battilani P. Studies on *Aspergillus* section Flavi isolated from maize in northern Italy. *Int J Food Microbiol* 2007;113:330–338.

[93] Cardwell KF, Cotty PJ. Distribution of *Aspergillus* Section Flavi among field soils from the four agroecological zones of the Republic of Benin, South Africa. *Plant Dis* 2002;86:434–439.

[94] Miller JD. Mycotoxins in small grains and maize: Old problems, new challenges. *Food Additives and Contaminants* 2008;25:219–230.

[95] Kriss AB, Paul PA, Madden LV. Relationship between yearly fluctuations in Fusarium head blight intensity and environmental variables: A window-pane analysis. *Phytopathology* 2010;100:784–797.

[96] Jones RK, Duncan HE, Hamilton PB. Planting date, harvest date, and irrigation effects on infection and aflatoxin production by *Aspergillus flavus* in field corn. *Phytopathology* 1981;71:810–816.

[97] Payne GA. *Aspergillus flavus* of maize silk and kernels. Aflatoxin in Maize. *Proceedings of the CIMMYT Workshop* 1986, Mexico. p. 119–126.

[98] Cotty PJ. Cottonseed losses and mycotoxins. In: Kirkpatrick TL, Rothrock CS, eds. *Compendium of Cotton Diseases*. Minnesota: The American Phytopathological Society; 2001. p. 9–13.

[99] Odvody GN, Spencer N, Remmers J. A description of silk cut, a stress-related loss of kernel integrity in preharvest maize. *Plant Dis* 1997;81:439–444.

[100] Salvacion AR, Ortiz BV, Scully BT, Wilson DM, Hoogenboom G, Lee RD. Effect of rainfall and maximum temperature on corn aflatoxin contamination in the Southeast US. *Climate Information for Managing Risks 2011,* May 24–27, 2011, Orlando, Florida.

[101] Del Ponte EM, Fernandes JMC, Pavan W, Baethgen WE. A model-based assessment of the impacts of climate variability on Fusarium head blight seasonal risk in southern Brazil. *J Phytopathol* 2009;157:675–681.

[102] Xu X, Nicholson P. Community ecology of fungal pathogens causing wheat head blight. *Ann Rev Phytopathol* 2009;47:83–103.

[103] Waalwijk C, van der Lee T, Yang L, de Vries I, Gortz A, Kerma G. Are changes in the composition of the Fusarium head blight complex caused by climate change? *Pests and Climate Change* 2008;240–241.

[104] Yli-Mattila T. Ecology and evolution oftoxigenic *Fusarium* species in cereals in northern Europe and Asia. *J Plant Pathol* 2010;92:7–18.

[105] Moore GG. Sex and recombination in aflatoxigenic Aspergilli: global implications. *Front Microb* 2014;5:1–5.

[106] Xu XM, Nicholson P, Thomsett MA, et al. Relationship between the fungal complex causing Fusarium head blight of wheat and environmental conditions. *Phytopathology* 2008;98: 69–78.

[107] Qu B, Li HP, Zhang JB, et al. Geographic distribution and genetic diversity of *Fusarium graminearum* and *F. asiaticum* on wheat spikes throughout China. *Plant Pathol* 2008;57: 15–24.

[108] Marasas WFO, Nelson PE, Toussoun TA. Toxigenic *Fusarium* Species: Identity and Mycotoxicology. University Park, PA: The Pennsylvania State University Press; 1984.

[109] Bottalico A. Fusarium diseases of cereals: species complex and related mycotoxins profiles in Europe. *J Plant Pathol* 1998;80:85–103.

[110] Logrieco A, Mule G, Moretti A, Bottalico A. Toxigenic *Fusarium* species and mycotoxins associated with maize ear rot in Europe. *Eur J Plant Pathol* 2002;108:597–609.

[111] Miller JD, Greenholgh R, Wang YZ, Lu M. Trichothecene chemotype of three *Fusarium* species. *Mycologia* 1991;83:121–130.

[112] Ward TJ, Clear RM, Rooney AP. An adaptive evolutionary shift in Fusarium head blight pathogen populations is driving the rapid spread of more toxigenic *Fusarium graminearum* in North Ameica. *Fungal Genet Biol* 2008;45:473–484.

[113] Vujanovic V, Goh YK, Daida P. Heat- and cold-shock responses in *Fusarium graminearum* 3 acetyl- and 15 acetyl-deoxynivalenol chemotypes. *The Journal of Microbiology* 2012;50: 97–102.

[114] Zhang H, Zhang Z, van der Lee T, et al. Population genetic analysis of *Fusarium asiaticum* populations from barley suggest a recent shift favoring 3ADON producers in southern China. *Phytopathology* 2010;100:328–336.

[115] Gilbert J, Brûlé-Babel A, Guerrieri AT, et al. Ratio of 3-ADON and 15-ADON isolates of *Fusarium graminearum* recovered from wheat kernels in Manitoba from 2008 to 2012. *Can J Plant Pathol* 2014;36:54–63.

[116] Spolti P, Del Ponte EM, Cummings AYD, Bergstrom GC. Fitness attributes of *Fusarium graminearum* isoaltes from wheat in New York possessing a 3-ADON or a 15-ADON tricothecene genotpye. *Phytopathology* 2014;104:513–519.

[117] von der Ohe C, Gauthier V, Tamburic-Ilincic L, et al. A comparison of aggressiveness and deoxynivalenol production between Canadian *Fusarium graminearum* isolates with 3-acetyl and 15-acetyldeoxynivalenol chemotypes in field-grown spring wheat. *Eur J Plant Pathol* 2010;127:407–417.

[118] Malihipour A, Gilbert J, Piercey-Normore M, Cloutier S. Molecular phylogenetic analysis, trichothecene chemotype patterns, and variation in aggressiveness of *Fusarium* isolates causing headblight in wheat. *Plant Dis* 2012;96:1016–1025.

[119] Szecsi A, Bartok T. Trichothecene chemotypes of *Fusarium graminearum* from corn in Hungary. *Mycotoxin Research* 1995;11:85–92.

[120] Cumagun CJR, Bowden RL, Jurgenson JJ, Leslie JF, Miedaner T. Genetic mapping of pathogenicity and aggressiveness of *Gibberella zeae* (*Fusarium graminearum*) toward wheat. *Phytopathology* 2004;94:520–526

[121] Cumagun CJR, Miedaner T. Segregation for aggressiveness and deoxynivalenol production of a population of *Gibberella zeae* (*Fusarium graminearum*). *Eur J Plant Pathol* 2004;110: 789–799.

[122] Cumagun CJR, Vargas MLD, Alviar AN. Evaluation of greenhouse and field aggressiveness of *Fusarium verticillioides* from corn in Laguna province, Philippines. *Afr J Agric Res* 2011;6:6586–6591.

[123] Probst C, Njapau H, Cotty PJ. Outbreak of an acute aflatoxicosis in Kenya in 2004: Identification of the causal agent. *Appl Environ Microb* 2007;73:2762–2764.

[124] Cotty PJ, Probst C, Jaime-Garcia R. Etiology and management of aflatoxin contamination. In: Leslie JF, Bandyopadhyay R, Visconti R, eds. *Mycotoxins, Detection Methods, Management, Public Health and Agricultural Trade*. Oxfordshire, UK: CAB International; 2008. p. 287–299.

[125] Probst C, Bandyopadhyay R, Cotty PJ. Diversity of aflatoxin-producing fungi and their impact on food safety in sub-Saharan Africa. *Int J Food Microbiol* 2014;174:113–122.

[126] Donner M, Atehnkeng J, Sikora RA, Bandyopadhyay R, Cotty PJ. Distribution of *Aspergillus* section Flavi in soils of maize fields in three agroecological zones of Nigeria. *Soil Biol Biochem* 2009;41:37–44.

[127] Teng PS, Savary S. Implementing the systems approach in pest management. *Agr Syst* 1992;40:237–264.

[128] Shaw M. Preparing for changes in plant disease due to climate change. *Plant Prot Sci* 2009;45:S3-S10.
[129] Ingram JSI, Gregory PJ, Izac A-M. The role of agronomic research in climate change and food security policy. *Agr Ecosys Environ* 2008;126:4–12.
[130] Guisan A, Zimmermann NE. Predictive habitat distribution models in ecology. *Ecol Model* 2000;135:147–186.
[131] Pearson RG, Dawson TP. Predicting the impacts of climate change on the distribution of species: are bioclimate envelope models useful? *Global Ecol Biogeogr* 2003;12:361–371.
[132] Sutherst RW, Constable F, Finlay KJ, Harrington R, Luck J, Zalucki MP. Adapting to crop pest and pathogen risks under climate change. *Wiley Interdisciplinary Reviews: Climate Change* 2011;2:220–237.
[133] Philippine Statistical Authority-Bureau of Agricultural Statistics (PSA-BAS). 2014 CountrySTAT Philippines. http://countrystat.bas.gov.ph. (accessed: 9 June 2014).
[134] Sumner PE, Lee D. Reducing aflatoxin in corn during harvest and storage. *The University of Georgia Cooperative Extension Bulletin* 2012.
[135] Stewart DW, Reid LM, Nicol RW, Schaafsma AW. A mathematical simulation of growth of *Fusarium* in maize ears after inoculation. *Phytopathology* 2002;92:534–541.
[136] R: A language and environment for statistical computing R Foundation for Statistical Computing 2014. at http://www.R-project.org/ (accessed 9 June 2014).
[137] Quitco RT. Aflatoxin studies in the Philippines. In: Champ BR, Highley E, Hocking AD, eds. *Fungi and Mycotoxins in Stored Products*. Bangkok, Thailand; 1991. p. 180–186.
[138] FAO. Aflatoxin contamination in foods and feeds in the Philippines. *FAO/WHO Regional Conference on Food Safety for Asia and Pacific*, 2004, Seremban, Malaysia, FAO.
[139] Cumagun CJR. Population genetic analysis of plant pathogenic fungi with emphasis on *Fusarium* species. *Philipp Agric Sci* 2007;90:244–256.
[140] Cumagun CJR, Ramos JS, Dimaano AO, Munaut F, Van Hove F. Genetic characteristics of *Fusarium verticillioides* from corn in the Philippines. *J Gen Plant Pathol* 2009;75:405–412.
[141] Programme WF. *Philippine Food and Nutrition Security Atlas*. World Food Programme – Philippines; 2012.
[142] Schaafsma AW, Hooker DC. Climatic models to predict occurrence of *Fusarium* toxins in wheat and maize. *Int J Food Microbiol* 2007;119:116–125.
[143] Boland GJ, Meizer MS, Hopkin A, Higgins V, Nassurth A. Climate change and plant diseases in Ontario. *Can J Plant Pathol* 2004;26:335–350.
[144] Madgwick JW, West JW, White RP, et al. Impacts of climate change on wheat anthesis and fusarium ear blight in the UK. *Eur J Plant Pathol* 2011;130:117–131.
[145] Roos J, Hopkins R, Kvarnheden A, Dixelius C. The impact of global warming on plant diseases and insect vectors in Sweden. *Eur J Plant Pathol* 2011;129:9–19.
[146] West JS, Townsend JA, Stevens M, Fitt BDL. Comparative biology of different plant pathogens to estimate effects of climate change on crop diseases in Europe. *Eur J Plant Pathol* 2012;133:315–331.
[147] van der Fels-Klerx HJ, Olesen JE, Madsen MS, Goedhart PW. Climate change increases deoxynivalenol contamination of wheat in north-western Europe. *Food Addit Contam* A 2012;29:1593–1604.
[148] Zhang X, Halder J, White RP, et al. Climate change increases risk of Fusarium ear blight on wheat in central China. *Ann Appl Biol* 2014;164:384–395.
[149] Van der Fels-Klerx HJ, van Asselt ED, Madsen MS, Olesen JE. Impact of climate change effects on contamination of cereal grains with deoxynivalenol. PLOS ONE 2013;1–6.
[150] Fernandes JM, Cunha GR, Del Ponte E, et al. Modeling Fusarium head blight in wheat under climate change using linked process-based models. *Proceedings of the 2nd International*

Symposium on Fusarium Head Blight, incorporating the 8th European Fusarium Seminar 2004: 11–15 December 2004: Orlando. p. 441–443.
[151] Battilani P, Rossi V, Giorni P, et al. *Modeling, predicting, and mapping the emergence of aflatoxins in cereals in the EU due to climate change*. Scientific report submitted to EFSA 2012.
[152] Juroszek P, Tiedemann AV. Climate change and potential future risks through wheat diseases: a review. *Eur J Plant Pathol* 2013;136:21–33.
[153] Shaw MW, Osborne TM. Geographic distribution of plant pathogens in response to climate change. *Plant Pathol* 2011;60:31–43.
[154] McIntyre B, Kim Y, Fruci J, Rosenthal E. The best-laid plans: climate change and food security. *Climate and Development* 2011;3:281–284.
[155] Garrett KA, Dobson ADM, Kroschel J, et al. The effects of climate variability and the color of weather time series on agricultural diseases and pests, and on decisions for their management. *Agr Forest Meteorol* 2013;170:216–227.
[156] Terzi V, Tumino G, Stanca AM, Morcia C. Reducing the incidence of cereal head infection and myctoxins in small grain cereal species. *Journal of Cereal Science* 2013;59:284–293.
[157] Pangga IB, Hanan J, Chakraborty S. Pathogen dynamics in a crop canopy and their evolution under changing climate. *Plant Pathol* 2011;60:70–81.
[158] Pangga IB, Hanan J, Chakraborty S. Climate change impacts on plant canopy architecture: implications for pest and pathogen management. *Eur J Plant Pathol* 2013;135:595–610.
[159] Willyerd KT, Li C, Madden LV, et al. Efficacy and stability of integrating fungicide and cultivar resistance to manage Fusarium head blight and deoxynivalenol in wheat. *Plant Dis* 2012;96:957–967.
[160] Thornton PK, Ericksen PJ, Herrero M, Challinor AJ. Climate variability and vulnerability to climate change: a review. *Glob Change Biol* 2014.
[161] Cruz A, Marin P, Gonzalez-Jaen MT, Aguilar KG, Cumagun CJR. Phylogenetic analysis, fumonisin production and pathogenicity of *Fusarium fujikuroi* strains isolated from rice in the Philippines. *J Sci Food Agric* 2013;93:3032–3039.
[162] Cumagun CJR. The rise of bakanae disease of rice in the Philippines. *Forum for Agriculture Risk Management in Development* 2012.
[163] Campos-Sánchez V, Esker PD, Murillo-Williams A. A shift in the mycotoxin profile in Costa Rica under climate change – A case study on mycotoxigenic *Fusarium proliferatum* in rice (*Oryza sativa*). *International Conference on Climate Change Impacts and Adaptation for Food and Environmental Security*. 21–22 November 2012, Conference Summary Report, Los Banos, Laguna, SEARCA. p. 23.
[164] Muller MEH, Brenning A, Verch G, Koszinski S, Sommer M. Multifactorial spatial analysis of mycotoxin contamination of winter wheat at the field and landscape scale. *Agr Ecosyst Environ* 2010;139:245–254.
[165] Marvin HJP, Kleter GA, Van der Fels-Klerx HJ, et al. Proactive systems for early warning of potential impacts of natural disasters on food safety: Climate change-induced extreme events a case in point. *Food Control* 2013;34:444–456.
[166] Gaudet DA, Tronsmo AM, Laroche A. Climate change and plant diseases. In: Storey KB, Tanino KK, eds. *Temperature Adaptation in a Changing Climate: Nature at Risk*. Wallingford: CAB International, 2012. p. 144–159.
[167] Dixon MA. Climate change – impacts on crop growth and food production, and plant pathogens. *Can J Plant Pathol* 2012;34:262–379.
[168] West JS, Holdgate S, Townsend JA, Edwards SG, Jennings P, Fitt BDL. Impacts of changing climate and agronomic factors on Fusarium ear blight of wheat in the UK. *Fungal Ecol* 2012;5:53–61.

X. Li and X. B. Yang

2 Impact of climate change on genetically engineered plants and mycotoxigenic fungi in the north central region of the US

2.1 Introduction

Global climate change is a major challenge facing modern human society [1]. The impact of climate change to aspects of human life and the global environment are profound and full of uncertainty. Climate change has great impact on ecosystems as well, especially agro-ecosystems [1–3]. In agriculture, plant diseases present a great threat to global food security [4]. Most economic losses are due to direct yield losses in the field and the cost of disease management. It has been estimated that diseases and other pests cause 42 % loss in the 8 most important food and cash crops worldwide [5].

A plant disease system belongs to an agro-ecosystem, which consists of the host, the pathogen(s), and the environment [6, 7]. The environment, largely associated with regional climate, interacts with both the host and the pathogen(s). Plant pathogens are usually very sensitive to weather conditions as well as climatic changes [6, 7]. Chakraborty et al. [8] suggested that climate change may have an impact on plant diseases in three main aspects: geographic distribution, intensity of disease occurrence, and effects of disease management. Since the 1980s, a number of researchers have addressed the impact of climate change on agriculture and plant diseases, and many of these studies were set in the US [1, 3, 8–16].

The US is one of the largest agricultural producers in the world in terms of planting area, yield, and economic value. Historically, a number of infectious crop diseases have attacked US agriculture, especially the row crops in the Great Plains and North Central region (Fig. 2.1). Many outbreaks were due to native or exotic fungal pathogens, e.g., southern corn leaf blight by *Cochliobolus heterostrophus*, corn gray leaf spot by *Cercospora zeae-maydis*, wheat rusts by *Puccinia* spp., wheat head blight by *Fusarium graminearum*, and soybean rust (exotic) by *Phakopsora pachyrhizi* [6, 17–24]. It has been estimated that plant diseases caused about $ 33 billion economic loss each year on average to US agriculture [5, 25].

A group of plant fungal diseases are caused by mycotoxigenic fungi, which are capable of producing secondary toxic substances in their metabolisms. Mycotoxigenic fungi are widely spread throughout the world and cause diseases in major food and row crops such as corn, cotton, soybeans, and wheat in many countries. These diseases not only cause yield losses, but also adversely impact on human health and society [6, 26, 27]. The damage from mycotoxigenic fungi may occur at different times or places, such as during crop production, postharvest/storage, food processing, or

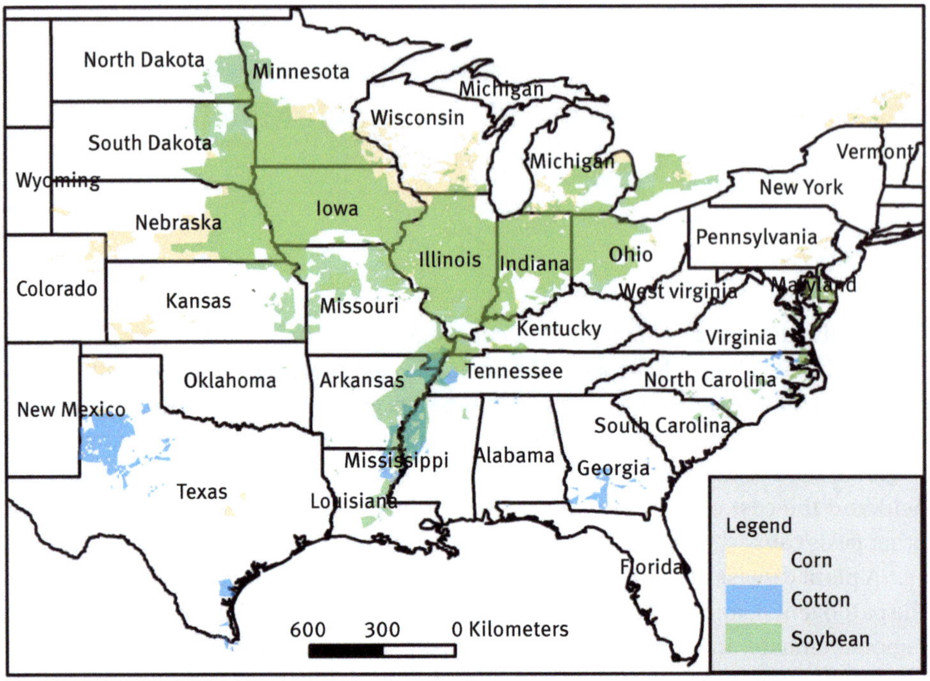

Fig. 2.1: Distribution of three field crops: corn, cotton, and soybeans in the US. The majority of these crops are genetically engineered varieties. Corn and soybeans are largely overlapping in the states of Illinois, Indiana, Iowa, and Ohio.

commodity trading. Yet the most important aspect is when the toxins produced by these fungi enter food and animal feed which may then cause various diseases, such as cancer, or acute mycotoxicoses in human and animals, sometimes fatal. For example, several aflatoxins are produced by *Aspergillus* spp. in different crops. Since their discovery in the early 1960s, aflatoxins have been considered among the most carcinogenic chemicals to human [27]. The discovery was largely due to the death of 100 000 turkeys after they fed on peanut meal contaminated by aflatoxins in London in 1962 [28]. An outbreak of mycotoxicosis caused by aflatoxins in Kenya in 2004 claimed 125 lives of 317 diagnosed cases [29]. Similar cases are common in developing countries where mycotoxigenic fungi are not properly managed and harvested grains are often stored under natural conditions [30–32]. Deoxynivalenol (DON) or vomitoxin produced by *F. graminearum* is another mycotoxin commonly detected in wheat, corn, and barley grains worldwide. When high dose DON from contaminated grains is ingested by animals it may cause acute nausea, vomiting, and diarrhea.

Similar to the situation in many countries, there are a variety of diseases caused by mycotoxigenic fungi in the US (Tab. 2.1). For instance, major toxins (aflatoxins from *Aspergillus* spp., fumonisin from *Fusarium* spp., and zearalenone from *Gib-*

Tab. 2.1: Important mycotoxigenic fungi and mycotoxins commonly found in major field crops in the United States.

Mycotoxigenic fungi	Mycotoxins produced	Affected genetically engineered crops	Favorable environments
Aspergillus flavus *Aspergillus parasiticus*	Aflatoxins (B1, B2, G1, G2, and M1)	Corn, cotton, rapeseed, soybeans	Hot and dry in growing season, humid at or after harvest
Fusarium verticillioides (*syn. F. moniliforme*) *Fusarium proliferatum*	Fumonisin	Corn	Warm and humid
Fusarium graminearum (teleomorph *Gibberella zeae*)	Deoxynivalenol (DON) Zearalenone	Corn	Warm, humid, and rainy

berella zeae) are commonly detected in the corn-growing regions. In wheat, DON from *F. graminearum*, which causes wheat scab or head blight, is of the most concern. Cotton, peanuts, and soybeans can be infected by *Aspergillus* spp., causing aflatoxin contaminations of lesser importance compared with the mycotoxin problems in corn and wheat. Several other less important mycotoxigenic fungi in fruits, nuts, and feed crops also occur in this country [26, 27, 33].

A number of major crops suffer significant losses from mycotoxins in the US, especially from aflatoxin [34–36]. Estimates of these losses in terms of economic value vary greatly. One estimated economic loss as a result of major mycotoxins, i.e. aflatoxins, fumonisin, and deoxynivalenol, in the US was about $ 932 million per year, plus an average of $ 466 million annually for regulatory enforcement, testing, and other quality control measures [35]. More specifically, aflatoxins caused an average $ 225 million economic loss in feed corn and $ 25.8 million loss in peanuts per year in the US from 1993 to 1996 [35, 37]. The estimated loss due to DON contamination in the US was $ 655 million per year, mostly in the wheat industry [38].

The presence of mycotoxins, particularly aflatoxins and DON, in food and feed is regulated by the Food and Drug Administration (FDA) to insure quality and safety in the US. A series of regulatory limits or action levels for aflatoxin of different products has been established by the FDA since the 1960s [39]. For instance, the aflatoxin action level is 20 ppb for all products for human consumption, except milk, and for all corn used for immature animals and dairy cattle. The action level for aflatoxin M1 in milk is only 0.5 ppb. The FDA currently sets up advisory levels for DON, e.g., 1 ppm in finished wheat products for human consumption. In grain production and trading, hundreds of grain samples were tested each year by elevators and exporters for mycotoxins. The threshold level, or so-called Limit of Detection (LOD) used by the US Grain Council is different for various mycotoxins. For example, in the 2013 season, the limits were 2.5 ppb for aflatoxins and 0.3 ppm for DON [40].

The occurrence of mycotoxigenic fungi is impacted by a changing climate too. Because of their importance in agriculture production, food safety, and public health, climate change effects on these diseases is a concern for researchers, the public, and governments worldwide [10, 32, 33, 38, 41–49]. However, in the US, the situation differs to some extent compared to other major agricultural regions.

The most notable characteristic in US agriculture regarding mycotoxigenic fungi is the high adoption rate of genetically modified (GM) crops. GM crops are genetically modified organisms (GMO) by definition. A GMO is obtained normally by genetic engineering (GE) or genetic modification techniques, which directly manipulate the DNA sequence in an organism's genome so that the modified genome will have modified or new genes to express new traits. In a GMO, specific genes may be inserted, modified, or deleted, so that the traits associated with these particular genes will express differently from the original organism. Using beneficial genes from other organisms, a GM crop can have good traits which cannot be normally obtained by classical breeding techniques, e.g., hybridization. The beneficial genes can be genes which resist herbicides, insects, or abiotic stresses, which produce useful byproducts, or improve the quality of nutrition and harvesting.

As a major advocator of the development of GMO techniques, the US has the longest history of using GM crops on a large scale. The US has the largest GM cropping system in the world, with 70.1 million hectares of GM crops planted in 2013 [50]. The GM crops widely used in this country include alfalfa, corn, cotton, papaya, rapeseeds, soybeans, squash, and sugar beet. Several others, such as GM wheat and rice, are ready for commercial release [50]. Currently, Roundup Ready (RR, glyphosate-resistant) and insect-resistant Bt (bacterium *Bacillus thuringiensis*) transgenic crops are the major ones, including RR alfalfa, corn, soybeans etc., and Bt cotton, corn, and potato. Therefore, the occurrence of diseases caused by mycotoxigenic fungi in GM crops in the US is largely different from that in other countries where GM crops were recently planted on a large scale, because GM crops have a great impact on these disease systems [51–58]. The effects of a changing climate on these fungal diseases are more complicated in the US than in other agricultural regions.

On the other hand, global climate change does not occur evenly in different regions. Greater changes, mainly warming and changes in precipitation patterns, occurred in the middle to high latitude regions of the northern hemisphere [1]. The major US agricultural zones occur in these regions. Therefore, the impacts of climate change on mycotoxigenic diseases in GM cropping systems in the US may be of greater magnitude and particularly important compared to other major GM crop planting countries, such as Brazil, because climate change has generally had less impact on tropical countries [1]. In this chapter we will focus on the important mycotoxigenic fungi of the major GM crops planted in the mid-latitude regions, i.e. the northern central US, and discuss the impact of climate change on the interactions and possible future trends.

2.2 GMO cropping systems in US agriculture

2.2.1 The establishment of GMO cropping systems in the US

The first GM plant was developed in 1982 by the Monsanto Company, which expressed an antibiotic-resistant gene in tobacco plants [59]. The first field tests of GM plants, with only a marker gene, were in the US and France in 1986 to develop a herbicide-resistant tobacco. The US then conducted 1952 field trials of GMO between 1986 and 1995 [60]. A large number of herbicide-resistant (or herbicide-tolerant), e.g. resistant to 2,4-D, glyphosate, and bromoxynil, GM crops were developed in response to the great need for weed management in agriculture. The first GM crop approved in the European Union was a herbicide-resistant tobacco in 1994 [60]. In the US, the glyphosate-resistant technique was developed by Monsanto Company and released first in the soybean market in 1996. Another major type of transgenic plant, insect-resistant plants, began in 1987, when an insecticidal protein gene *cry* from *B. thuringiensis* (Bt) was transferred to tobacco plants by Plant Genetic Systems, previously a Belgian company [56]. This technique was soon employed in cotton, corn, potato, and other crops once developed.

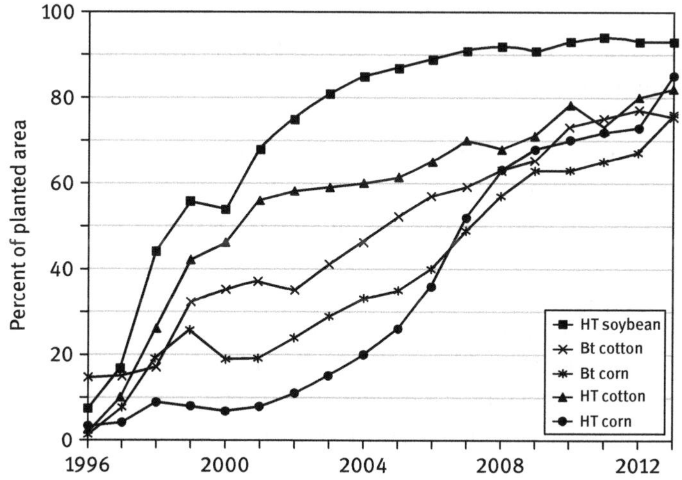

Fig. 2.2: Adoption of genetically engineered crops in the United States 1996–2013. HT: herbicide tolerant traits. Bt: *Bacillus thuringiensis* toxin traits.

From 1996 the use of glyphosate-resistant and Bt techniques soared in soybeans, corn, and cotton in the US (Fig. 2.2). By 2013, about 80–90 % of corn, cotton, and soybeans in the US were GM crops. The distribution of these GM crops covers a vast geographic area (Fig. 2.1), including the Corn Belt and the Great Plains. Glyphosate-resistant and Bt crops are the most successfully used GM crops worldwide. About 90 % of them (in terms of planting area; millions of hectares) are concentrated in 5 countries: the US (70.1), Brazil (40.3), Argentina (24.4), India (11), and Canada (10.8), mainly in the canola, corn, cotton, and soybean cropping systems [50].

2.2.2 Glyphosate-resistant crops and Bt transgenic techniques

Glyphosate (N-(phosphonomethyl) glycine), the active ingredient in Roundup® products, was introduced to the market by Monsanto in 1971 as a non-selective and systemic herbicide [61]. It targets an enzyme (5-enolpyruvylshikimic acid-3-phosphate (EPSP) synthase) commonly found in plants, fungi, and bacteria. Glyphosate inhibits this enzyme and causes reduced protein synthesis, growth, and premature cell death in sensitive organisms [62]. Glyphosate has low toxicity to human and animals with a Toxicity Class of III (on a scale of I to IV, IV is the least toxic) according to the US Environmental Protection Agency's (EPA) assessment. This makes it safer than many other commonly used herbicides, such as paraquat and imazapyr. Thus, it has been widely used in agriculture, horticulture, and home gardening. Roundup Ready® (RR) refers to a GM crop with a glyphosate resistant trait, which enables the GM crop to resist glyphosate in culturing practices [63]. It allows producers to use glyphosate against a wide spectrum of weeds without killing the crops. Glyphosate resistance genes have been licensed to several major US seed companies, such as Pioneer and Syngenta, by Monsanto to be used in alfalfa, corn, soybeans, and wheat (RR wheat is not currently on the market).

Bt crops are based on the transformation of toxin-encoding genes from *B. thuringiensis* to crops to create insect-resistant traits [64]. *B. thuringiensis* itself can be used as a biocontrol agent in the field. There are a number of *cry* genes (Cry1A.105, CryIAb, CryIF, Cry2Ab, Cry3Bb1, Cry34Ab1, Cry35Ab1, and mCry3A) which can produce delta endotoxins to kill lepidopteran and coleopteran larvae, especially European corn borer (*Ostrinia nubilalis*) and western corn rootworm (*Diabrotica virgifera virgifera*) larvae [64]. They can be used either individually or stacked in a GM crop. Another class of Bt genes used against insects is vegetative insecticidal protein (VIP) genes, which have a different mode of action from *cry* genes [65, 66].

2.2.3 General impact of GM crops on US agriculture

GM crops have had a profound impact on US agriculture [50, 51, 54, 58, 63, 64, 67–73]. Breeders have developed crops with stacked transgenic traits, e.g. corn varieties with Bt and RR traits. This has allowed growers to reduce applications of insecticide and other herbicides. According to estimations by Benbrook [67], using Bt transgenic crops reduced insecticide applications in the field by about 56 000 tons during 1996–2011. Inevitably, the use of glyphosate increased drastically [74]. It was estimated that glyphosate use in soybean fields increased from 3935 tons in 1996 to 40 273 tons in 2006 (Chemical Use Program, National Agricultural Statistics Service (NASS), United States Department of Agriculture (USDA)). Along with roundup ready technique, post-emergence weed management became simpler, more flexible, and often cheaper [69, 75]. However, after two decades of intensive use of glyphosate in GM cropping systems, more and more weed populations, such as waterhemp (*Amaranthus rudis*), horseweed (*Conyza canadensis*), and ragweed (*Ambrosia artemisiifolia and A. trifida*) have evolved high resistance to this herbicide, and are sometimes called super weeds [67, 70, 76]. In response to the glyphosate-tolerant weeds, several other herbicide-resistant GM crops have been developed. These new herbicide-resistant crops include Enlist®, incorporating both glyphosate resistance and 2,4-D resistance, and LibertyLink®, resistant to Liberty® herbicide (Glufosinate ammonium), but their market share is still much smaller than RR crops. Similar to glyphosate-resistant crops, Bt crops also put steady selection pressure on the target insects, resulting in the emergence of insect populations resistant to the *cry* and VIP traits [77]. Collateral damage to other non-targeted insects and GMO pollen pollution of some organic farming systems may occur as well.

The use of herbicide-tolerant, mainly glyphosate-resistant, crops helped the adoption of conservation tillage and no-till in the US greatly [68]. According to a report by the USDA, no-till in corn, cotton, soybeans, and rice increased at a median rate of about 1.5 % per year from 2000–2007 [72]. The soybean-corn rotation systems in the corn belt had the highest increase, with about 50 % no-till in soybean and 29.5 % in corn crops in 2009. Also, a large percentage of fields is under conservation tillage, in which large amounts (usually > 30 %) of crop residue are left on the soil surface. No-till reduces soil erosion and compaction with less operational costs in labor and machinery, improves water and fertilizer use efficiency, and creates less disturbed habitats for biocontrol agents and beneficial insect predators [54, 73, 78]. On the other hand, increased crop residues seem to have contributed to recent epidemics of several diseases, e.g., Fusarium head blight in wheat (*F. graminearum*) and gray leaf spot in corn (*C. zeae-maydis*) [17, 18, 79]. Also, surface plant residues reduce soil surface temperatures and temperature fluctuation, and provide a better habitat for microorganisms, some of which are pathogenic [51, 54].

2.2.4 Impact of GM crops on mycotoxigenic fungi

Using GM crops to reduce the damage of mycotoxigenic fungi is also one of the purposes of developing genetic modification techniques [80]. In Bt GM cropping systems, particularly Bt corn, crop injuries resulting from insects, e.g. European corn borer, were significantly reduced, leading to reduced *Fusarium* spp. infections and mycotoxin concentration in kernels [55, 81]. The estimated direct economic benefits based on the use of Bt crops against mycotoxigenic fungi was about $ 22 million annually according to an analysis by Wu [58] in 2004, without considering potential benefits in aspects of human and animal health [57]. On the other hand, no-till or reduced tillage increases plant residue on the soil surface significantly, providing more food sources for the survival of several major mycotoxigenic fungi. The recent increases in Fusarium head blight of wheat and ear/stalk rot of corn were attributed to this by various researchers [17, 79, 82, 83]. Recently, serious debate on the intense use of glyphosate in RR cropping systems and its impact on plant diseases has been initiated by researchers who suggested that glyphosate has caused increases in certain diseases [52, 53, 84, 85]. Most of their evidence is from *Fusarium* spp., including some mycotoxigenic ones [86–88]. The mechanisms are not fully understood yet. It is likely that glyphosate changes the balance of soil microenvironment, nutrition situation, and competition among fungi in soil, especially around the rhizosphere [53, 71].

2.3 Global climate change and current situation in the US

Climate change on global or regional scales has been discussed in numerous research articles, books, and reports [1, 2, 9, 10, 89–91]. In the latest assessment report (AR5) by the Intergovernmental Panel on Climate Change (IPCC), the major trends of global climate change have been further clarified. According to various investigations and observational data, average global surface temperature has increased by 0.85 °C (0.65 to 1.06 °C) since 1880, along with diminished glaciers and rising sea levels. On a climatic scale, the last 30 years have been the warmest period since 1850 based on human instrumental records, and are also likely to have been the warmest in the northern hemisphere in the last 1400 years [1]. Cold days are fewer and warm days are more frequent globally, resulting in an earlier spring [92]. Changes in patterns of extreme weather and climate events have been observed in recent decades in many regions [1].

Climate change differs in various aspects on a regional scale. More heat waves seem to have occurred in Australia, Asia, and Europe. In the mid-latitude areas of the northern hemisphere, precipitation has increased since 1951 with high statistical confidence. In North America and Europe, heavy rainfall likely increased in more land regions, though drought prevailed in some areas [1, 89, 90]. Observations of extreme weather events, such as tropical cyclones, increased in several regions [91].

The 2012 growing season was the warmest in the US since 1895 according to the US National Climatic Data Center (NCDC). In most regions of the US, annual mean temperatures have increased in general, though significant geographical differences exist. For example, Alaska, which is at high latitude, had the largest increase, followed by the northern Midwest, i.e., north central region, and the southwest. Seasonally, autumn temperatures increased the least and winter temperatures increased the most. Warmer springs were seen in most regions as well, e.g. the warmest on record in 2012. The warming in spring has resulted in longer growing season, at a changing rate of

1–4 days longer per decade [10]. In Iowa, a north central state, the annual number of frost-free days increased by 8–9 days from 1893–2008 [9]. Changes in summer temperatures were more complex: in the southeast they were actually cooling, although the trend is diminishing, while the north central region had slightly warmer summers [10, 90]. Annually, there is generally more rainfall now in the northwestern, central, and southern US than 100 years ago. The most notable rainfall increase occurred in central US. According to observations in the north central region, where mainly corn and soybeans are grown, rainfall, mainly in spring and summer, increased by about 15 % in the last 100 years [9, 89]. The southern US had more rainfall in autumn [10, 90].

The increase in precipitation has two aspects: frequency and intensity. Generally, there are more rainy days, and more heavy rains. Such an increase cooled the region, sometimes called the "warming hole", in the US north central compared to the temperature increases on the whole continent [93, 94]. An increase in summer rainfall can lead to a wet and warm growing season, such as those of 2010 and 2011, or a wet and cool growing season, such as the 2009 season, which was close to the coolest summer recorded in North Central (Data from NCDC). In contrast, parts of the eastern US, the Rocky Mountains, and much of the southwest received less rainfall, and southwestern US experienced more frequent drought. Most recently, severe droughts occurred in 2012 and 2013 in the southern Great Plains, where wheat is the major crop, as well as part of the north central region, resulting in yield losses of corn (Data from DCDC).

In the US, heavy rainfall, drought, heat waves, tornados, and hurricanes are weather extremes observed in growing seasons. Increases in the frequency of weather extremes are considered important indicators for changes in regional climates. With increased occurrences of weather extremes, annual variations of weather patterns may become greater than before [10, 89, 90]. According to NCDC data, extreme weather events have increased in frequency in recent years, causing significantly different weather patterns each year, particularly in the north central region, for example the record cool summer in 2009, the very wet summer in 2010 with record floods in several states, a hot and humid 2011 summer leading to outbreaks of fungal diseases in corn and soybeans, and a sudden severe drought in 2012. The 2013 season was of no difference with a near record wettest May and cool/dry summer in the north central region (DCDC). An extreme event in one year is insufficient to suggest climate change in a region. When extreme weather patterns occur several years in a row, we can confidently say that this region is experiencing significant changes in climate.

2.4 Impacts of climate change on the occurrence of mycotoxigenic fungi

2.4.1 The impact of climate change in the off-seasons: winter and early spring

Most fungi are very sensitive to changes in environment, including mycotoxigenic ones. Among the major mycotoxigenic fungi in the US, *Penicillium* spp are of less concern because they are more important during storage than in the growing season, and the US has advanced storage facilities to minimize their impact. During the growing season, *Fusarium* spp. and *Aspergillus* spp. (Tab. 2.1), the major fungal agents causing disease and toxin contamination, would certainly be impacted more by a changing climate.

Generally speaking, these diseases have relatively simple disease cycles. These fungi normally overwinter either as mycelium or as resistant structures, e.g., chlamydospores (*F. graminearum*), thickened hyphae (*F. verticillioides*), or as sclerotia (*Aspergillus* spp.) in soil or on crop residues, mainly corn [95–97]. As mentioned above, warming in winter and spring were most significant in temperate climate zones, especially in middle to high latitude regions, such as the north central region in the US [1, 9, 90]. Winter temperatures in these locations can frequently reach –10 to –20 °C or even lower. *Fusarium* spp. in crop residue have long-term survivability and high tolerance of cold weather. A study conducted in Kansas by Manzo and Claflin [98] assessing the survival rate of *F. moniliforme* hyphae and conidia in sorghum residue showed that the fungus had no loss of viability after 6 months at –16 °C. Cotten and Munkvold [99] reported that when buried in soil in Iowa, nearly half of the corn residue inoculated with *F. moniliforme, F. proliferatum, and F. subglutinans* still had the fungi one year later, similar to survival rates on the surface. Inch and Gilbert [100] found that *G. zeae* could survive in wheat kernels for at least 2 years on soil surface in Canada. These reports suggest that winter warming in cold regions would have minimal impact on the survival of *Fusarium* spp. in crop residue.

Aspergillus species prefer a warmer environment, thus winter temperatures would certainly appear more important for their survival in the north. It is well-documented that sclerotia of *Aspergillus* species are abundant in the southern US, especially *A. flavus* [101]. Even in southern Texas, a warm region, the density of *A. flavus* propagule followed a negative correlation in soil daily temperature range of 18–30 °C across different seasons [102]. In the north, e.g. Iowa, low levels of *A. flavus* presence could still be detected in years of unfavorable conditions with a high annual variation in population density [103]. But Shearer et al [104] suggested that these fungi do not produce sclerotia in the north as readily as in the southern US due to lower autumn temperatures in September and October which are unfavorable for sclerotia development. A comparative study by Wicklow [97] showed that the decay of conidia of *Aspergillus* species was slower in Illinois than that in Georgia, and *A. parasiticus* survived better than *A. flavus*. It seems that low winter temperatures in the north may favor survival of conidia of *Aspergillus* species, but cool autumn weather still hinders the production of sclerotia, the longer-surviving fruiting body. Warming in winter and early spring in the north may impair the survival of local *Aspergillus* species in the long-run since their population seems to decline faster in a warmer environment, while unfavorable environment conditions for propagule production in summer and autumn do not change much according to observed climate changes. In addition, various bacteria and fungi may colonize sclerotia of *Aspergillus* species, reducing their viability [97, 105]. Warming in winter and spring may enhance these parasitic activities, deteriorating the soil environment for *Aspergillus* sclerotia.

Insects which are vectors of these pathogens are certainly affected by climate change in winter as well [106]. A warmer winter and spring allow more generations a year and a gradual expansion of the distribution range of the European corn borer and corn rootworms further north in the US and Canada has been predicted [107, 108]. In the newly colonized areas in the north, low population of these insects would perhaps not have much impact on diseases caused by *Aspergillus* spp., since low summer temperatures are more important as a limiting factor to infections and to mycotoxin contamination than insect damage. In the Corn Belt, the European corn borer population normally produces two generations in a season or part of the 3rd generation in an extended growing season. Their overwintering populations generally suffer less cold stress in warm winters if only the air temperature increases [109, 110]. However, warming in winter also leads to more rain in winter in some areas, such as Iowa [9, 10]. If the warming causes reduced snow cover or more frequent freeze-thaw cycles during the period of insect diapause, higher mortality may occur for some insects, e.g. larvae of the European corn borer and corn earworm (*Heliothis zea*), which overwinter better in a snow-covered or less moist environment [106, 111, 112]. Nonetheless, compared to the effects of Bt corn which is widely used in this region, impacts of climate change in winter may be secondary in affecting insects and related fungal diseases.

2.4.2 Impact of climate change on planting date and fungi at seedling stages

Planting dates are important to the seasonal occurrence of many crop diseases in the US. Earlier planting has been adopted by many growers, largely due to the earlier occurrence of spring. For example, corn can be planted in mid-April in Iowa, the core planting area in the US Corn Belt (Fig. 2.3). As Fusarium ear rot grows better in a moderate temperature range, warm weather early in a spring would be highly favorable to the pathogen's activities in surface crop residue, leading to more propagule production. However, the temperature below the soil surface does not increase rapidly in fields with no-till or conservation tillage because of the crop residue coverage [73]. Thus, seedlings in early planting fields may suffer from longer exposure to low soil temperatures, a relatively stressful condition. Observations in recent years from Ohio State suggested that seedling diseases caused by *F. graminearum* and other *Fusarium* spp.

in wheat-corn-soybean rotation appeared increased even after seeds were treated with fungicides [113]. For the early seedling stage, conidia on residues in the soil may be the major sources of inoculum since they can overwinter as well [100]. The increase of *Fusarium* seedlings may be related to the increased activities of Fusaria under warmer conditions in conjunction with increased inoculum from infested crop residue.

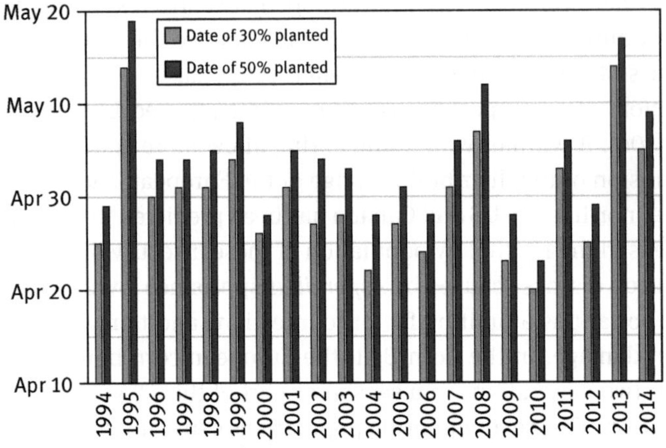

Fig. 2.3: Planting dates of corn in Iowa in the last 20 years according to the weekly crop progress reports issued by the United States Department of Agriculture. There have been more early plantings since the 2000s, though planting was delayed in two recent years due to excessively wet springs. A similar trend can be found in soybean planting, which normally occurs after corn planting.

Early planting allows crops to accumulate biomass for a longer period. Increased biomass in no-till or conservation tillage is favorable to the initial inoculum production in crop residue, which certainly favors the survival of mycotoxigenic fungi on crop residues, especially *Fusarium* spp. since they do not produce sclerotia. On the other hand, in the US Corn Belt, 2nd generation European corn borers usually cause more damage, especially in late planted corn [114]. Early planted corn or early maturing corn may escape this period of great damage, having less damaged kernels for the infection of mycotoxigenic fungi. However, this effect may also be secondary and could be masked by the suppressive effect of Bt toxins on insect populations.

Fast warming in early spring causes rapid melting of snow and sometimes heavy convective precipitation during the early planting season [10]. A wetter spring causes fewer workable days in the field, leading to planting difficulties [9]. For example, in the 2013 spring, Iowa experienced an extended planting season delay, and many fields were planted in June, after the wettest May on record (NCDC). According to the Crop Progress Reports released by NASS of USDA, by May 19, only 71 % of corn fields were planted, much less compared to the average level of 92 % in the previous 5 years. Soybeans are normally planted after corn. By June 2, only about 44 % of the soybean fields in Iowa had been planted, in contrast to the average level of the previous 5 years, which was 91 % by that date. Late planted corn suffers the greatest damage from 2nd generation European corn borers. This in turn is favorable for the occurrence of *F. verticillioides, F. proliferatum, or F. subglutinans*, the dominant species for Fusarium ear rot via insect wounds [115–118], leading to fumonisin contamination if warm and dry weather persists during the late reproductive stages. In another case, later maturing fields may escape the infection peak of *F. graminearum* causing Gibberella ear rot during the silking stage, which is normally in July. However, cool and wet conditions in late summer and early autumn are favorable for *F. graminearum* infection and DON contamination in kernels if severe insect damage has occurred.

Even with normal or early planting, heavy spring rainfall can cause field to flood. A few days of flooding can also slow crop growth, reduce crop stands, or delay harvest. In extreme cases, impact may be passed on to the next season because crops in a late planted field may not be harvested due to early frost or other weather events during the harvest season. Consequently, massive amounts of infected kernels and crop residue are left on the surface for the next season as primary inoculum sources. Collectively, frequent rainfall in spring involves more uncertainty regarding disease occurrence.

2.4.3 Impact of climate change on crops and diseases in late spring and summer

Late spring and summer are the key transition stages for crop development from the vegetative to the reproductive stages. Weather events will have great impact on inoculum dispersal, infection, and development of mycotoxigenic diseases. In the north central US, changes in temperatures in late spring and summer were minor, but precipitation pattern changes were significant, resulting in heavy storms, floods, drought, and crop physical injuries [9, 89]. As mentioned above, events of heavy rainfalls increased in the north central US. In Iowa (Fig. 2.4), the Storm Events Database maintained by NCDC indicate that events of damaging heavy rain and storm wind (strong winds arising from convection with speeds > 50 knots, i.e. 92.6 kph) have increased in recent decades in terms of overall occurrence, affected areas, occurrence time, and strength. These events lead to great uncertainty in disease occurrence.

Several major mycotoxigenic fungi, particularly *A. flavus* and *F. graminearum*, infect kernels during the silking stages of corn, a critical time for disease development. The primary sources of inoculum for *Fusarium* spp. and *Aspergillus* spp. infections are from fungal structures on residue and other crop tissues [82, 83, 95, 118, 119]. For instance, infectious propagules of *F. graminearum* are ascospores discharged from perithecia, survival structures in residue [95], and macroconidia in sporodochia in crop residue may also serve as a minor source [83]. Rainfall has a strong influence on the initial infection of *F. graminearum* because ascospore discharge normally occurs when perithecia are dried during the day and moistened suddenly at night, either by high relative humidity or light rain [120] with 16 °C as optimal [121]. A rainfall event

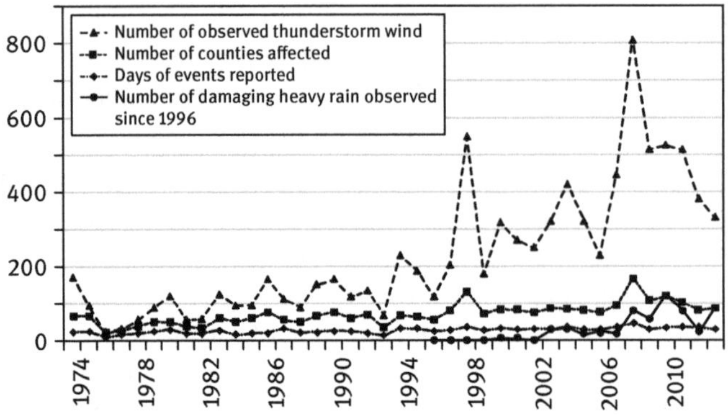

Fig. 2.4: Numbers of strong storm winds and damaging heavy rains observed May–August in Iowa in the north central US region in recent decades.

can easily have suitable environmental conditions for ascospore discharge by providing moisture and low temperature. Thus, the observed increase of rainy days and humidity in the north central region during the corn silking stage brings more frequent changes in air moisture and cool weather [9], thus favoring discharge and infection of *F. graminearum* ascospores, setting the initial disease level high [83].

Corn kernel infections by *F. verticillioides, F. subglutinans, and F. proliferatum* are mainly via airborne microconidia and macroconidia on crop residues [83]. The sclerotia of *Aspergillus* spp.germinate to produce hyphae or conidia, which can be dispersed in the soil or carried by winds and insects to infect corn [119]. *A. parasiticus* infects mainly peanut, while *A. flavus* appears more on corn and cotton, where aerial infections mainly occur, and is thus more important in the north central region [119, 123, 124]. The more frequent occurrence of heavy rainfall and high speed storm winds favor dispersal and wet deposition of these airborne inocula among fields and within crop canopies [125]. Thus, a higher prevalence of Gibberella and Fusarium ear rot in this region could occur.

Extreme precipitation causes various crop stressors, such as flood, hail damage, and lodging. Stressed crops are less resistant to fungal infections, particularly to those entering through wounds. *Fusarium* and *Aspergillus* species can infect plants through wounds, thus weather extremes such as heavy rainfall, hail storms, strong winds will increase infections. However, disease occurrence data associated with these events are scarce. Another aspect of the impact of heavy precipitation could be on insect development. Normally, a dry year is more favorable for insects. Heavy rainfall may cause high mortality during insect egg hatching. For *F. verticillioides*, frequent heavy rainfalls could reduce European corn borer infestations, which is important in spore dispersal [126]. Cool weather during a period of rainy days may hinder development of insect larvae, leading to lower population density in the late season.

Warming is not as significant as that in winter and spring in the north central region during late spring and summer. This region is a temperate zone, thus days with high temperatures above 30 °C are relatively rare in states such as Iowa, Illinois, and Indiana. Yet, increased night temperatures have been observed in this region [9, 90]. It has been clear that warming in the night impairs corn yield greatly [127–129]. For example, the hot summer nights of 2011 cut the national corn yield to 147.2 bushels per acre (NASS/USDA). Furthermore, elevated temperatures often cause heat stress in corn in the early reproductive stages, but less so in soybeans, if drought occurs in the meantime. A recent case was in 2012: average corn yield dropped to 123.4 bushels per acre due to hot and dry weather (NASS/ USDA).

In general, *Fusarium* spp. prefer a moderate temperature range, though *F. graminearum* prefers cooler temperatures (optimal about 24–28 °C), while *F. verticillioides* and the others prefer the warmer side (optimal about 30 °C) [130–132]. They therefore prevail in the central to northern US. A hot season will be unfavorable to Gibberella ear rot. For instance, the testing results for DON reported by the US Grain Council showed that 91.6 % of samples had less than 0.5 ppm of the mycotoxin in 2013, slightly lower than in 2012 at 96 %, while the 2012 summer was much warmer (about 2–4 °C regionally) than 2013 in the north central region [40]. Overall, current changes in summer temperatures may not impact Fusarium and Gibberella ear rot in this region as much as other factors such as no-till and heavy precipitations.

In contrast, *A. flavus* favors hot and dry conditions. The fungus can grow well in a temperature range of 25–42 °C, but favors a high optimum temperature of 37 °C [119]. High night temperatures, above 30 °C, certainly favor kernel infection of *A. flavus* [119, 133]. Insects are also favored by hot and dry weather, resulting in rapid development of new generations and causing more kernel wounds. These aspects of negative impacts from hot and dry weather would thus create a synergetic effect for *Aspergillus* ear rot, leading to rapid disease progress and mycotoxin contamination in a hot and dry year. Historically, when much warmer than normal and relatively dry summers occurred in Iowa, high rates of aflatoxin contamination occurred in grain, such as in 1983 (37 % positive) and 1988 (30 % positive) [134]. According to harvest quality reports from the US Grain Council, in 2012, the percentage of samples with no detectable aflatoxin was 78 %, much lower than that in 2013 (about 98 %) [40], indicating the impact of hot dry weather in 2012 on the occurrence of *A. flavus*. In 2012, in response to the severe drought and potential aflatoxin contamination, the state government of Iowa requested mandatory testing of milk for aflatoxins in late summer [135]. Although these data were not systematically collected, given that July of 2012 was much hotter and drier than July of 1983, the lower rates of aflatoxin-positive samples in 2012 (22 %) than in the previous drought (1983) may reflect the significant effect of Bt GM crops on controlling mycotoxin contamination. Nonetheless, regional drought and hot weather similar to that in 2012 may alert to potential problems of mycotoxin contamination resulting from extreme weather events from climate change in the future, not mentioning that insects are also evolving with higher resistance to Bt toxins [77].

2.4.4 Increased use of fungicides

Fungicide use in row crops such as corn and soybeans has increased significantly in the US Corn Belt since the year 2000, including seed treatment and foliar spray [136–138]. There are several reasons for this increase: increasing crop prices making fungicides more affordable, more disease pressure resulting from early planting and no-till practices, new fungicides such as strobilurin available for these crops, favorable environments due to climate change, and even the invasion of continental US with soybean rust [139]. Many researchers still suggest that fungicide spray should be applied only when disease pressure is high [138–141]. In corn, it was reported that fungicide application can reduce Fusarium stalk rot significantly, leading to higher yield by reducing lodging from stalk rot. Meanwhile, seed treatment has been widely used in corn and soybean fields to protect seeds from seedling disease in the early season, helping adoption of early planting. These changes are partially under the influence of climate change in this region and would inevitably affect mycotoxigenic fungi. Naturally, fungicides can protect crops from mycotoxigenic fungal infection, leading to reduced occurrence of mycotoxin contamination.

2.5 Summary and future risks

The impact of climate change in north central US on local mycotoxigenic fungi is great, and directly or indirectly affects the regional occurrence of diseases caused by mycotoxin-producing fungi. Directly, climate change affects fungal survival, growth, dispersal, infection, and disease development. Indirectly, climate change affects planting dates, culturing practices, and disease management, etc. The impacts of climate change are different for different crop diseases. Current changes in climate favor the occurrence of Gibberella ear rot in corn because of increased rainfall, warmer spring weather, and more crop residue. More occurrence of this disease is likely to be seen in the Corn Belt as in wheat production regions such as Indiana and Ohio. The occurrence of Fusarium ear rot in this region may increase slightly in the near future as well because frequent thunderstorms aid their dispersal and initial infections, as well as slight summer warming. Bt crops and fungicides will still play important roles in reducing the threat. *Aspergillus* spp. is severe in the north central US when severe hot dry summer weather similar to that in 2012 occurs, especially for more than one year in a row.

However, these factors combined with the use of GM crops, will have synergistic effects on mycotoxigenic fungi, leading to greater uncertainty and difficulties in predicting future trends [142]. It is even more difficult to make a long-term prediction for disease risk due to the lack of monitoring data and comprehensive understanding of the often highly non-linear synergistic impacts [143]. Nonetheless, it is clear that the regional climate in north central US has experienced more weather extremes and greater annual fluctuations in recent decades. Under these circumstances, it is likely to have seasons which meet the occurrence conditions of the major mycotoxigenic fungi in the near future. This implies that the local diversity and prevalence of the diseases caused by mycotoxigenic fungi may increase, though the overall severity, except Gibberella ear rot, may not necessarily be much higher than before, especially when the insects are well controlled by Bt crops and fungicides are used. In the long run, if the regional temperatures increase as predicted by IPCC and extended droughts occur, particularly in summer and autumn, *Aspergillus* spp. could become more a significant threat.

References

[1] IPCC. *Climate change 2013: the physical science basis.* Contribution of Working Group I to the Fifth Assessment Report of the Intergovernmental Panel on Climate Change. Cambridge, United Kingdom and New York, NY, USA: Cambridge University Press; 2013.

[2] IPCC. *Climate change 2007: the physical science basis.* Contribution of Working Group I to the Fourth Assessment Report of the Intergovernmental Panel on Climate Change. Cambridge, United Kingdom and New York, NY, USA: Cambridge University Press; 2007.

[3] Rosenzweig C, Iglesias A, Yang XB, Epstein PR, Chivian E. Climate change and extreme weather events; implications for food production, plant diseases, and pests. *Global change & human health* 2001;2:90–104.

[4] Strange RN, Scott PR. Plant disease: A threat to global food Security, *Ann Rev Phytopathol* 2005;43:83–116.

[5] Oerke EC, Dehne HW, Schonbeck F, Weber A. *Crop production and crop protection: Estimated losses in major food and cash crops.* Amsterdam, the Netherlands: Elsevier; 1994.

[6] Agrios G. *Plant Pathology.* Burlington, MA, USA: Academic Press; 2005.

[7] Kranz J. *Epidemics of plant diseases.* New York, NY, USA: Springer Verlag; 1990.

[8] Chakraborty S, Tiedemann AV, Teng PS. Climate change: potential impact on plant diseases. *Environmental Pollution* 2000;108:317–326.

[9] Iowa Climate Change Impacts Committee. *Climate change impacts on Iowa 2010: Report to the Governor and the Iowa General Assembly.* Iowa Climate Change Impacts Committee, Des Moines, IA, USA; 2010.

[10] Walthall C, Hatfield J, Backlund P, et al. *Climate change and agriculture in the United States: Effects and adaptation.* USDA Technical Bulletin 1935. Washington DC, USA; 2012.

[11] Anderson PK, Cunningham AA, Patel NG, Morales FJ, Epstein PR, Daszak P. Emerging infectious diseases of plants: Pathogen pollution, climate change and agro-technology drivers. *Trends Ecol Evol* 2004;19:535–544.

[12] Coakley SM, Scherm H. Plant disease in a changing global environment. *Aspects Appl Biol* 1996;45:227–238.

[13] Coakley SM, Scherm H, Chakraborty S. Climate change and plant disease management. *Ann Rev Phytopathol* 1999;37:399–426.

[14] Garrett KA, Dendy SP, Frank EE, Rouse MN, Travers SE. Climate change effects on plant disease: Genomes to ecosystems. *Ann Rev Phytopathol* 2006;44:489–509.

[15] Garrett KA, Dobson ADM, Kroschel J, et al. The effects of climate variability and the color of weather time series on agricultural diseases and pests, and on decisions for their management. *Agricultural and Forest Meteorology* 2013;170:216–227.

[16] Chakraborty S, Newton AC. Climate change, plant diseases and food security: An overview. *Plant Pathology* 2011;60:2–14.
[17] McMullen M, Roger J, Gallenberg D. Scab of wheat and barley: a re-emerging disease of devastating impact. *Plant disease* 1997;81:1340–1348.
[18] Ward JM, Stromberg EL, Nowell DC, Nutter FW Jr. Gray leaf spot: A disease of global importance in maize production. *Plant Disease* 1999;83:884–895.
[19] Ullstrup AJ. The impacts of the southern corn leaf blight epidemics of 1970–1971. *Ann Rev Phytopathol* 1972;10:37–50.
[20] Li X, Esker PD, Pan Z, Dias AP, Xue L, Yang XB. The uniqueness of the soybean rust pathosystem: An improved understanding of the risk in different regions of the world. *Plant Disease* 2010;94:796–806.
[21] Roelfs AP, Bushnell WR. *The Cereal Rusts*. Vol. II, New York, NY, USA: Academic Press; 1985.
[22] McMullen M, Bergstrom G, De Wolf E, et al. A unified effort to fight an enemy of wheat and barley: Fusarium head blight. *Plant Disease* 2012;96:1712–1728.
[23] Tatum LA. The southern corn leaf blight epidemic. *Science* 1971;171:1113–1116.
[24] Brown JKM, Hovmøller MS. Aerial dispersal of pathogens on the global and continental scales and its impact on plant disease. *Science* 2002;297:537–541.
[25] Pimentel D, Lach L, Zuniga R, Morrison D. Environmental and economic costs of nonindigenous species in the United States. *Bioscience* 2000;50:53–65.
[26] Richard JL. Some major mycotoxins and their mycotoxicoses–an overview. *International journal of food microbiology* 2007;119:3–10.
[27] Bennett JW, Klich MA. Mycotoxins. *Clinical Microbiology Reviews* 2003;106:497–516.
[28] Van der Zijden AS, Koelensmid WAA, Boldingh J. *Aspergillus flavus* and Turkey X disease: Isolation in crystalline form of a toxin responsible for Turkey X disease. *Nature* 1962;195:1060–1062.
[29] Azziz-Baumgartner E, Lindblade K, Gieseker K, et al. Case-control study of an acute aflatoxicosis outbreak in Kenya, 2004. *Environmental Health Perspectives* 2004;113:1779–1783.
[30] Williams JH, Phillips TD, Jolly PE, Stiles JK, Jolly CM, Aggarwal D. Human aflatoxicosis in developing countries: a review of toxicology, exposure, potential health consequences, and interventions. *The American Journal of Clinical Nutrition* 2004;80:1106–1122.
[31] Ngindu A, Johnson BK, Kenya PR, Ngira JA, Ocheng DM. Outbreak of acute hepatitis caused by aflatoxin poisoning in Kenya. *Lancet* 1982;319:1346–1348.
[32] Shephard GS. Impact of mycotoxins on human health in developing countries. *Food Additives and Contaminants* 2008;25:146–151.
[33] Wood GE. Mycotoxins in foods and feeds in the United States. *Journal of Animal Science* 1992;70:3941–3949.
[34] Robens J, Cardwell K. The costs of mycotoxin management to the USA: Management of aflatoxins in the United States. *Journal of Toxicology* 2003;22:139–152.
[35] CAST. Mycotoxins: Risks in plant, animal, and human systems. Ames, IA, USA: Council of Agricultural Science and Technology; 2003.
[36] Zubera MS, Lillehoj EB. Status of the aflatoxin problem in Corn. *J Environ Qual* 1979;8:1–5.
[37] Lamb MC, Sternitzke DA. Cost of aflatoxin to the farmer, buying point, and sheller segments of the Southeast United States peanut industry. *Peanut Science* 2001;28:59–63.
[38] Schmale DG, Munkvold GP. Mycotoxins in crops: *A threat to human and domestic animal health*. In: The Plant Health Instructor. St. Paul, MN, USA: American Phytopathological Society; 2009 (accessed June 20, 2014, at http://www.apsnet.org/edcenter/intropp/topics/mycotoxins/Pages/default.aspx.)

[39] NGFA. *FDA mycotoxin regulatory guidance: A guide for grain elevators, feed manufacturers, grain processors and exporters*. Washington DC, USA: National Grain and Feed Association; 2011.
[40] U.S. Grains Council. *Corn export cargo quality report 2013/14*. Washington DC, USA: U.S. Grains Council; 2014.
[41] Scholthof KBG. One foot in the furrow: Linkages between agriculture, plant pathology, and public health. *Annu Rev Public Health* 2003;24:153–74.
[42] Tirado MC, Clarke R, Jaykus LA, McQuatters-Gollop A, Frank JM. Climate change and food safety: A review. *Food Research International* 2010;43:1745–1765.
[43] Miraglia M, Marvin HJP, Kleter GA, et al. Climate change and food safety: an emerging issue with special focus on Europe. *Food and Chemical Toxicology* 2009;47:1009–1021.
[44] Lake IR, Hooper L, Abdelhamid A, et al. Climate change and food security: Health impacts in developed countries. *Environmental Health Perspectives* 2012;120:1520–1526.
[45] Wu F, Bhatnagar D, Bui-Klimke T, et al. Climate change impacts on mycotoxin risks in US maize. *World Mycotoxin Journal* 2011;4:79–93.
[46] Schaafsma AW, Hooker DC. Climatic models to predict occurrence of *Fusarium* toxins in wheat and maize. *International Journal of Food Microbiology* 2007;119:116–125.
[47] Paterson RRM, Lima N. Further mycotoxin effects from climate change. *Food Research International* 2011;44:2555–2566.
[48] Paterson RRM, Lima N. How will climate change affect mycotoxins in food? *Food Research International* 2010;43:1902–1914.
[49] Cotty PJ, Jaime-Garcia R. Influences of climate on aflatoxin producing fungi and aflatoxin contamination. *International Journal of Food Microbiology* 2007;119:109–115.
[50] James C. *Global status of commercialized biotech/GM crops: 2013*. Ithaca, New York, NY, USA: International Service For the Acquisition of Agri-biotech Applications; 2013.
[51] Dunfield KE, Germida JJ. Impact of genetically modified crops on soil- and plant-associated microbial communities. *J Eviron Qual* 2004;33:806–815.
[52] Johal GS, Huber DM. Glyphosate effects on disease and disease resistance in plants. *Eur J Agron* 2009;31:144–152.
[53] Kremer RJ, Means NE. Glyphosate and glyphosate-resistant crop interactions with rhizosphere microorganisms. *Eur J Agron* 2009;31:153–161.
[54] Lundgren JG, Gassmann AJ, Bernal JJ, Duan JJ, Ruberson J. Ecological compatibility of GM crops and biological control. *Crop Protection* 2009;28:1017–1030.
[55] Munkvold GP, Hellmich RL, Rice LG. Comparison of fumonisin concentrations in kernels of transgenic Bt maize hybrids and nontransgenic hybrids. *Plant Disease* 1999;83:130–138.
[56] Vaeck M, Reynaerts A, Höfte H, et al. Transgenic plants protected from insect attack. *Nature* 1987;328:33–37.
[57] Wu F, Miller JD, Casman EA. The economic impact of Bt corn resulting from mycotoxin reduction. *Journal of Toxicology–Toxin Reviews* 2004;23:397–424.
[58] Wu F. Mycotoxin reduction in Bt corn: potential economic, health, and regulatory impacts. *Transgenic Research* 2006;15:277–289.
[59] Fraley RT, Rogers SG, Horsch RB, et al. Expression of bacterial genes in plant cells. *Proc Natl Acad Sci USA* 1983;80:4803–4807.
[60] James C, Krattiger AF. *Global review of the field testing and commercialization of transgenic plants: 1986 to 1995*. Ithaca, New York, NY, USA: The International Service for the Acquisition of Agri-biotech Applications; 1996.
[61] Franz JE, Mao MK, Sikorski JA. *Glyphosate: A unique global herbicide*. ACS Monograph No. 189. Washington DC, USA: American Chemical Society; 1997.

[62] Amrhein N, Deus B, Gehrke P, Steinrücken HC. The site of the inhibition of the shikimate pathway by glyphosate II. Interference of glyphosate with chorismate formation in vivo and in vitro. *Plant Physiology* 1980;66:830–834.
[63] Dill GM. Glyphosate-resistant crops: History, status. *Pest Manag Sci* 2005;61:219–224.
[64] Castagnola AS, Jurat-Fuentes JL. Chapter 15: Bt crops: past and future. In: Sansinenea E, (ed.) *Bacillus Thuringiensis Biotechnology*. New York, NY, USA: Springer Netherlands; 2012. p. 283–304.
[65] Yu CG, Mullins MA, Warren GW, Koziel MG, Estruch JJ. The *Bacillus thuringiensis* vegetative insecticidal protein Vip3A lyses midgut epithelium cells of susceptible insects. *Appl Environ Microbiol* 1997;63:532–536.
[66] Lee MK, Walters FS, Hart H, Palekar N, Chen JS. The mode of action of the *Bacillus thuringiensis* vegetative insecticidal protein Vip3A differs from that of Cry1Ab δ-Endotoxin. *Appl Environ Microbiol* 2003;69:4648–4657.
[67] Benbrook CM. Impacts of genetically engineered crops on pesticide use in the U.S. – the first sixteen years. *Environmental Sciences Europe* 2012;24:1–13.
[68] Cerdeira AL, Duke SO. The current status and environmental impacts of glyphosate-sesistant crops: A review. *J Environ Qual* 2006;35:1633–1658.
[69] Gianessi LP. Economic and herbicide use impacts of glyphosate-resistant crops. *Pest Manag Sci* 2005;61:241–245.
[70] Powles SB. Evolved glyphosate-resistant weeds around the world: Lessons to be learnt. *Pest Manag Sci* 2008;64:360–365.
[71] Zablotowicz, RM, Reddy KN. Impact of glyphosate on the *Bradyrhizobium japonicum* symbiosis with glyphosate-resistant transgenic soybean: A minireview. *J Environ Qual* 2004;33: 825–831.
[72] Horowitz J, Ebel R, Ueda K. *"No-Till" Farming Is a Growing Practice/EIB-70*. Washington DC, USA: Economic Research Service, United States Department of Agriculture; 2010.
[73] Triplett GB, Dick WA. No-tillage crop production: A revolution in agriculture!. *Agronomy Journal 100 Supplement 3* 2008;S-153.
[74] Shaner DL. The impact of glyphosate-tolerant crops on the use of other herbicides and on resistance management. *Pest Manag Sci* 2000;56:320–326.
[75] Wilcut SD, Askew JW. Cost and weed management with herbicide programs in glyphosate-resistant cotton (*Gossypium hirsutum*). *Weed Technology* 1999;13:308–313.
[76] Johnson WG, Davis V, Kruger G, Weller S. Influence of glyphosate-resistant cropping systems on weed species shifts and glyphosate-resistant weed populations. *Eur J Agron* 2009;31: 162–172.
[77] Sanchis V, Bourguet D. *Bacillus thuringiensis*: Applications in agriculture and insect resistance management: A review. *Agronomy for Sustainable Development* 2008;28:11–20.
[78] Holland JM. The environmental consequences of adopting conservation tillage in Europe: Reviewing the evidence. *Agriculture, Ecosystems & Environment* 2004;103:1–25.
[79] Dill-Macky R, Jones RK. The effect of previous crop residues and tillage on Fusarium head blight of wheat. *Plant Disease* 2000;84:71–76.
[80] Munkvold GP. Cultural and genetic approaches to manage mycotoxins in maize. *Ann Rev Phytopathol* 2003;41:99–116.
[81] Munkvold GP, Hellmich RL, Showers WB. Reduced Fusarium ear rot and symptomless infection in kernels of maize genetically engineered for European corn borer resistance. *Phytopathology* 1997;87:1071–1077.
[82] Parry D, Jenkinson P, McLeod L. Fusarium ear blight (scab) in small grain cereals – a review. *Plant Pathology* 1995;44:207–238.

[83] Munkvold GP. Epidemiology of *Fusarium* diseases and their mycotoxins in maize ears. *Eur J Plant Path* 2003;109:705–713.
[84] Fernandez MR, Zentner RP, Basnyat P, Gehl D, Selles F, Huber D. Glyphosate associations with cereal diseases caused by *Fusarium* spp. in the Canadian Prairies. *Eur J Agron* 2009;31: 133–143.
[85] Fernandez MR, Zentner RP, DePauw RM, Gehl D, Stevenson FC. Impacts of crop production factors on common root rot of barley in Eastern Saskatchewon. *Crop Sci* 2007;47:1585–1595.
[86] Lévesque CA, Rahe JE, Eaves DM. Effects of glyphosate on *Fusarium* spp.: Its influence on root colonization of weeds, propagule density in the soil, and crop emergence. *Can J Microbiol* 1987;33:354–360.
[87] Sanogo S, Yang XB, Scherm H. Effects of herbicides on *Fusarium solani* f. sp. *glycines* and development of sudden death syndrome in glyphosate-tolerant soybean. *Phytopathology* 2000;90:57–66.
[88] Fernandez MR, Selles F, Gehl D, DePauw RM, Zentner RP. Crop production factors associated with Fusarium head blight in spring wheat in eastern Saskatchewan. *Crop Sci* 2005;45: 1908–1916.
[89] Karl TR, Melillo JM, Peterson TC. *Global climate change impacts in the United States*. U.S. Global Climate Change Research Program. Cambirdge, United Kingdom: Cambridge University Press; 2009.
[90] Mearns L. *The North American regional climate change assessment program (NARCCAP): Overview of phase II results, 21st Conference on climate variability and change, Jan. 11–14, 2009, Phoenix, NV.*
[91] Webster PJ, Holland GJ, Curry JA, Chang HR. Changes in tropical cyclone number, duration, and intensity in a warming environment. *Nature* 2005;309:1844–1846.
[92] Stine AR, Huybers P, Fung IY. Changes in the phase of the annual cycle of surface temperature. *Nature* 2009;457:435–440.
[93] Kunkel KE, Liang XZ, Zhu J, Lin Y. Can CGCMs simulate the 20th century "warming hole" in the central United States. *J Climate* 2006;19:4137–4153.
[94] Pan Z, Arritt RW, Takle ES, Gutowski WJ Jr, Anderson CJ, Segal M. Altered hydrologic feedback in a warming climate introduces a "warming hole". *Geophysical Research Letters* 2004;31:L17109.
[95] Sutton JC. Epidemiology of wheat head blight and maize ear rot caused by *Fusarium graminearum*. *Can J Plant Path* 1982;4:195–209.
[96] Nyvall RF, Kommedahl T. Individual thickened hyphae as survival structures of *Fusarium moniliforme* in corn. *Phytopathology* 1968;58:1704–1707.
[97] Wicklow DT, Wilson DM, Nelsen TC. Survival of *Aspergillus flavus* sclerotia and conidia buried in soil in Illinois or Georgia. *Phytopathology* 1993;83:1141–1147.
[98] Manzo SK, Claflin LE. Survival of *Fusarium moniliforme* hyphae and conidia in grain sorghum stalks. *Plant Disease* 1984;68:866–867.
[99] Cotton TK, Munkvold GP. Survival of *Fusarium moniliforme*, *F. proliferatum*, and *F. subglutinans* in maize stalk residue. *Phytopathology* 1998;88:550–555.
[100] Inch SA, Gilbert J. Survival of *Gibberella zeae* in *Fusarium*-damaged wheat kernels. *Plant Disease* 2003;87:282–287.
[101] Born BW. Biodiversity of *Aspergillus* section *Flavi* in the United States: A review. *Food Additives and Contaminants* 2007;24:1088–1101.
[102] Jaime-Garcia R, Cotty PJ. Crop rotation and soil temperature influence the community structure of *Aspergillus flavus* in soil. *Soil Biology & Biochemistry* 2010;42:1842–1847.

[103] McGee DC, Olanya OM, Hoyos GM, Tiffany LH. Populations of *Aspergillus flavus* in the Iowa cornfield ecosystem in years not favorable for aflatoxin contamination of corn grain. *Plant Disease* 1996;80:742–746.

[104] Shearer JF, Sweets LE, Baker NK, Tiffany LH. A study of *Aspergillus flavus/parasiticus* in Iowa crop fields 1988–1990. *Plant Disease* 1992;76:19–22.

[105] Will ME, Wilson DM, Wicklow DT. Evaluation of *Paecilomyces lilacinus*, chitin, and cellulose amendments in the biological control of *Aspergillus flavus* fungi. *Biology and Fertility of Soils* 1994;17:281–284.

[106] Bale JS, Hayward SAL. Insect overwintering in a changing climate. *Journal of Experimental Biology* 2010;213:980–994.

[107] Diffenbaugh NS, Krupke CH, White MA, Alexander CE. Global warming presents new challenges for maize pest management. *Environment Research Letters* 2008;3:1–9.

[108] Kocmánková E, Trnka M, Eitzinger J, et al. Estimating the impact of climate change on the occurrence of selected pests at a high spatial resolution: A novel approach. *J Agri Sci* 2011;149:185–195.

[109] Bradshaw W, Holzapfel C. Evolutionary response to rapid climate change. *Science* 2006;312:1477–1478.

[110] Chiang HC, Hodson AC. Population fluctuations of the European corn borer, *Ostrinia nubilalis*, at Waseca, Minnesota, 1948–70. *Environmental Entomology* 1972;1:7–16.

[111] Barber GW, Dicke FF. Effect of temperature and moisture on overwintering pupae of the corn Earworm in the northeastern states. *J Agri Res* 1939;59(10):711–723.

[112] Hanec W, Beck SD. Cold hardiness in the European corn borer, *Pyrausta nubilalis* (Hübn.). *Journal of Insect Physiology* 1960;5:169–180.

[113] Broders KD, Lipps PE, Paul PA, Dorrance AE. Evaluation of *Fusarium graminearum* associated with corn and soybean seed and seedling disease in Ohio. *Plant Disease* 2007;91:1155–1160.

[114] Spangler SM, Calvin DD, Russo J, Schlegel J. Predicting risk of European corn borer infestation in sweet corn based on harvest date. HortTechnology 2009;19:173–180.

[115] Koehler B. *Corn ear rots in Illinois* Bulletin 639, Urbana-Champaign, IL, USA: Agricultural Experiment Station, University of Illinois; 1959.

[116] Kommedahl T, Windels CE. Root-, stalk-, and ear-infecting *Fusarium* species on corn in the USA. In: Nelson PE, Toussoun TA, Cook RJ (eds). *Fusarium: Diseases, Biology, and Taxonomy.* University Park, PA, USA: Pennsylvania State University Press; 1981. p. 94–104.

[117] Foley D. Systemic infection of corn by *Fusarium moniliforme. Phytopathology* 1962;52: 870–872.

[118] Smith DR, White DG. Diseases of corn. In: *Corn and Corn Improvement*, 3rd ed Agronomy Series No. 18. Am Soc Agronomy: Madison, WI, USA; 1988. p. 687–766.

[119] Diener UL, Cole RJ, Sanders TH, Payne GA, Lee LS, Klich MA. Epidemiology of aflatoxin formation by *Aspergillus flavus. Ann Rev Phytopathol* 1987;25:249–70.

[120] Paulitz T. Diurnal release of ascospores by *Gibberella zeae* in inoculated wheat plots. *Plant Disease* 1996;80:674–678.

[121] Tschanz AT, Horst RK, Nelson P E. The effect of environment on sexual reproduction of *Gibberella zeae. Mycologia* 1976;68:327–340.

[122] Aylor DE. Aerobiology of fungi in relation to capture and release by plants. In: Lindow SE, Hecht-Poinar EI, Elliott VJ (eds.) *Phyllosphere Microbiology.* American Phytopathological Society: St. Paul, MN, USA; 2002. p. 341–361.

[123] Diener VL, Davis ND. Invasion of peanut pods in the soil by *Aspergillus flavus. Plant Dis Rep* 1965;49:931–935.

[124] Gourama H, Bullerman LB. *Aspergillus flavus* and *Aspergillus parasiticus*: Aflatoxigenic fungi of concern in foods and feeds: A review. *Journal of Food Protection* 1995;58:1395–1404.

[125] Gregory PH. *The microbiology of the atmosphere*. John Wiley & Sons: New York, USA; 1973.
[126] Sobek EA, Munkvold GP. European corn borer larvae as vectors of *Fusarium moniliforme*, causing kernel rot and symptomless infection of maize kernels. *Journal of Economic Entomology* 1999;92:503–509.
[127] Peters D, Pendleton J, Hageman R, Brown CM. Effect of night air temperature on grain yield of corn, wheat, and soybeans. *Agron J* 1971;63:809.
[128] Chang JH. Corn yield in relation to photoperiod, night temperature, and solar radiation. *Agricultural Meteorology* 1981;24:253–262.
[129] Lobell DB, Asner GP. Climate and management contributions to recent trends in U.S. agricultural yields. *Science* 2003;299:1032.
[130] Reid LM, Nicol RW, Ouellet T, et al. Interaction of *Fusarium graminearum* and *F. moniliforme* in maize ears: Disease progress, fungal biomass, and mycotoxin accumulation. *Phytopathology* 1999;89:1028–1037.
[131] Booth C. The genus *Fusarium*. Kew, Surrey, England: Commonwealth Mycological Institute: 1971.
[132] Marín S, Magan N, Bellí N, Ramos AJ, Canela R, Sanchis V. Two-dimensional profiles of fumonisin B1 production by *Fusarium moniliforme* and *Fusarium proliferatum* in relation to environmental factors and potential for modelling toxin formation in maize grain. *International Journal of Food Microbiology* 1999;51:159–167.
[133] Jones RK, Duncan HE, Payne GA, Leonard KJ. Factors influencing infection by *Aspergillus flavus* in silk-inoculated corn. *Plant Disease* 1980;64:859–863.
[134] Robens JF. *A perspective on aflatoxins in field crops and animal food products in the United States: A symposium*. ARS-83. United States Department of Agriculture, Agricultural Research Service (USDA-ARS): Peoria, Illinois, United States; 1990.
[135] Ingwersen J. *Iowa begins mandatory testing of milk for aflatoxin*, 2012 (accessed April 4, 2014, at http://www.reuters.com/article/2012/08/31/us-usa-drought-aflatoxin-iowa-idUSBRE87U0ZV20120831.)
[136] Hershman DE, Vincelli P, Kaiser CA. *Foliar fungicide use in corn and soybeans, PPFS-MISC-05*. Kentucky Cooperative Extension Service, University of Kentucky: Lexington, KY, USA; 2011.
[137] Dorrance AE, Cruz C, Mills D, et al. Effect of foliar fungicide and insecticide applications on soybeans in Ohio. Online. *Plant Health Progress* doi:10.1094/PHP-2010-0122-01-RS, 2010.
[138] Henry RS, Johnson WG, Wise KA. The impact of a fungicide and an insecticide on soybean growth, yield, and profitability. *Crop Protection* 2011;30:1629–1634.
[139] Wise K, Mueller D. *Are fungicides no longer just for fungi? An analysis of foliar fungicide use in corn*. American Phytopathological Society, St. Paul, MN, USA: APSnet Features. doi:10.1094/APSnetFeature-2011-0531; 2011.
[140] Paul PA, Madden LV, Bradley CA, et al. Meta-analysis of yield response of hybrid field corn to foliar fungicides in the U.S. Corn Belt. *Phytopathology* 2011;101:1122–1132.
[141] Swoboda C, Pedersen P. Effect of fungicide on soybean growth and yield. *Agron J* 2009; 101:352–356.
[142] Garrett KA, Forbes GA, Savary S, et al. Complexity in climate-change impacts: an analytical framework for effects mediated by plant disease. *Plant Pathology* 2011;60:15–30.
[143] Jeger MJ, Pautasso M. Plant disease and global change-the importance of long-term data sets. *New Phytologist* 2008,177:8–11. doi:10.1111/j.1469-8137.2007.02312.x

José-Miguel Barea

3 Interactions among plants, arbuscular mycorrhizal and mycotoxigenic fungi related to food crop health in a scenario of climate change

3.1 Introduction

Climate change is a widely accepted probability, therefore its impacts on agro-ecosystem stability and productivity must be considered [1–3]. Predictive climate models support the likelihood that temperature increases and drought will have negative influences on agricultural productivity in the present century [2]. Diverse studies have therefore been performed to assess the impact of climate change on the different agents and scenarios which interact in relation to agricultural developments [3]. One field which has attracted remarkable research interest is the potential repercussion of climate change on the incidence of plant diseases [4–6], with special emphasis on those affecting production and quality of the main staple food crops as they can negatively affect food security, especially in resource-poor regions [7, 8].

The 2009 Declaration of the World Summit on Food Security [9] considered that "food security exists when all people, at all times, have physical, social and economic access to sufficient, safe and nutritious food which meets their dietary needs and food preferences for an active and healthy life", identifying four pillars of food security: availability, access, utilization, and stability. Agricultural output, especially cereal crop production, will have to increase by 70 % to feed a world population expected to surpass 9 billion in 2050 [9]. Cereals constitute approximately 56 % and 44 % of the human and animal, energy consumption worldwide, respectively, and to meet demands by 2050 the cereal crop production is projected to increase from the 2.5 billion tons expected for 2014 to three billion tons [10, 11].

Diseases on cereals caused by species of the fungal genera *Fusarium* and *Aspergillus* are of particular concern because these fungi can not only cause important yield losses but also produce mycotoxins which enter the food chain and are detrimental to human and animal health [12]. Prior to harvest, the effects of climate change on mycotoxins may occur via the fungi, hosts, or host-fungal interactions [13]. A number of predictive models have been developed worldwide to estimate the risk of mycotoxin contamination in cereals which take variables related to mycotoxigenic fungi and plants into consideration[14], but none have taken the fact that most globally important food crops establish beneficial relationships with mutualistic fungi, which can help reduce crop losses from both biotic and abiotic above-belowground stresses and address issues of food security into account [15].

The multitrophic interactions which take place in soil, where the components of soil microbiome are significant protagonists, have a great influence on food crop production and health. These interactions are responsible for C sequestering, nutrient cycling, pathogen suppression, biotic and abiotic stress alleviation, etc. A review of the related literature suggests that climate change clearly affects these interactions, and the design of appropriate predictive modeling approaches has therefore been recommended. However, elaboration of these models is hampered by high levels of uncertainty when trying to put all the interacting agents together [3].

Among the members of plant-associated microbiomes, the arbuscular mycorrhizal (AM) fungi are recognized as one of the most beneficial groups of soil biota [16]. These fungi are known to establish mutualistic AM symbiosis with most globally important food crops, being fundamental for plant growth and health [17]. All cereals hosting mycotoxigenic fungi are known to form AM symbiosis, but information on interactions between AM and mycotoxigenic fungi is scarce.

The AM association can be considered an adaptive strategy which increases the plant's ability to capture nutrients and induces increased tolerance of environmental stress factors, either biotic (e.g. pathogen attack), or abiotic (e.g. drought, salinity, heavy metals, organic pollutants, etc.) [18]. In particular, AM symbiosis protects plants against deleterious organisms including microbial pathogens, herbivorous insects, and parasitic plants. It has recently been clearly established that AM colonization can prime plant immunity by boosting the plant's ability to respond to a pathogen attack, as part of the "mycorrhiza-induced systemic resistance" [19]. The diversity and activity of AM fungi, and the formation and functioning of AM symbiosis are all affected by climatic change [20–22].

Considering the above ideas, the aims of this chapter are: (1) to offer an overview of a key activity within the interactions between AM fungi and plants, namely the impact of AM symbiosis on plant stress alleviation, particularly in inducing resistance/tolerance of plants to/of pathogens and pests. Emphasis is put on the ability of AM symbiosis to control shoot pathogens by eliciting a systemic, plant-mediated resistance response, and the way in which AM colonization can prime plant immunity by boosting the plant's ability to respond to a pathogen attack, as part of the "mycorrhiza-induced systemic resistance" abilities of AM fungi. (2) To analyze the scarce publications on the interactions between mycotoxigenic and AM fungi in relation to crop health, and to propose appropriate research approaches to improve our understanding of these interactions; and (3) to discuss the perspectives, challenges, and opportunities for exploiting AM services to control the activities of mycotoxigenic fungi, with regards to promotion of food security in the current scenario of climatic change.

3.2 Arbuscular mycorrhizal (AM) symbiosis

Most plants growing on earth live associated with endophytic soil fungi forming the so-called mycorrhizas, but arbuscular mycorrhizal (AM) associations, the most common mycorrhizal type, form nearly 80 % of the major plant families, including economically important crop species [16].

AM fungi are universally distributed soil-borne microbial fungi which originated and diverged nearly 500 million years ago, as determined by molecular analyses and fossil records. They form AM symbioses, the result of a common evolution history with plants dating back more than 400 million years which seems to have facilitated the adaptation of plants to the terrestrial environment. Actually, there is evidence to suggest that AM fungi played a crucial role in land colonization by early plants and in their evolution [23].

AM fungi are a subset of endophyte microorganisms which colonize plant tissues symbiotically. Other endophytes include both bacteria and fungi, either mutualistic or pathogens [14, 24–27]. AM fungi are asexual, unculturable and obligatorily biotrophic microbes as they are unable to complete their life cycle without colonizing a host plant, a characteristic which hampered research of their biology and their biotechnological applications [28]. The AM fungi belong to the phylum *Glomeromycota* [23, 29, 30]. Diverse studies based on molecular approaches have indicated that individual AM fungal ecotypes possess little host specificity, but a certain degree of host preference (functional compatibility) was shown to occur [17].

Obviously, the more relevant biological characteristic of AM fungi is their capacity to form AM associations involving almost all phyla of land plants in all soil and biomes [16].

3.2.1 AM establishment, function, and management

AM fungi colonize plant roots from three types of propagules: spores, fragments of mycorrhizal roots, and extraradical hyphae, all of which are a mycelial network expanding in the soil. When a hypha from an AM mycelium approaches a host root, a molecular dialogue between both symbionts occurs. This activates specific signaling pathways which affect both fungal development and plant gene expression from appressorium formation to the intracellular accommodation of the fungal symbiont to ends with the formation of the characteristic tree-like structures, termed “arbuscules”, within the root cortical cells [28]. The arbuscules which give their name to the mycorrhizal type are involved in the nutrient exchange between fungus and plant. The major nutrient flux is the transfer of carbon from plant to fungus (and thereby to the soil), and the movement of mineral nutrients from fungus to plant.

Once AM fungi consolidate the internal colonization of plant roots, they form an extensive mycelial network outside the root by establishing an extraradical mycelium where AM spores develop, thereby ending the fungal life cycle. The extraradical mycelium is actually a tri-dimensional structure specialized in the acquisition of plant nutrients from soil, particularly those present in low concentrations in the soil solution or which diffuse slowly to the uptake zone at the root (the rhizosphere), as is the case with phosphate and ammonia [28]. In addition to the uptake of nutrients, AM symbiosis improves plant fitness by increasing resistance to/tolerance of environmental stresses, whether biotic (e.g. pathogens and pests) or abiotic (e.g. nutrient deficiencies, drought, salinity, heavy metals toxicity, or the presence of organic pollutants), and also enhances soil structure through the formation of hydro-stable aggregates necessary for good soil structure [28].

AM function concerning plant nutrient acquisition is based on the fact that the external mycelium of AM fungi extends from the roots to absorb and transport nutrients, at a distance of up to 25 cm, while hyphal length densities in field soils range from 3 to 14 m/g [16]. Analysis of the biochemical and molecular mechanisms involved in the coordinated nutrient transport processes in AM and in the bidirectional nutrient exchange between symbionts at the symbiotic interfaces is matter of recent and current interest, as recently reviewed [28]. In summary, transport proteins putatively involved in Pi uptake from soil solution by AM fungi and those implicated in the uptake of the Pi exported across the membrane of the arbuscule have been identified in several plant species. Gene expression studies of the plant Pi transporters have revealed that development of symbiosis induces the *novo* expression of the so-called mycorrhiza-specific Pi transporters, which are exclusively expressed in mycorrhizal roots, and up-regulation of Pi transporters which have a basal expression in non-mycorrhizal roots, the mycorrhiza up-regulated transporters.

Some advances have also been made in recent years in understanding the mechanisms of N transport. The N taken up by extraradical hyphae (either as ammonium or nitrate), is assimilated in the fungal cytoplasm into arginine, transferred via the tubular vacuoles to the intraradical hyphae to be released as urea, and either transported to the plant directly or after cleavage as ammonium. Genes encoding ammonium transporters have so far been identified [31]. Both profiling technologies and gene-to-phenotype reverse genetic tools are revealing the plant and fungal protein involved in sustaining AM symbiosis, mostly on the basis of host-induced gene silencing approaches [32].

AM fungi management has attracted greater interest as a result of the increased demand for low-input agriculture because the manipulation and use of beneficial soil microorganisms is recognized as a recommended strategy for more sustainable food crop production based on the reduction of agrochemical maintenance while minimizing environmental degradation [17, 33]. Because of the importance of AM symbiosis in sustainable agriculture, the development of techniques for AM inoculant production and inoculation has become a central research topic and the available information was recently reviewed [18]. The difficulty in culturing obligate symbionts such as AM fungi in the absence of their host plant is a major obstacle for massive production of inoculum. Despite this problem, a number of companies worldwide are producing AM inoculum products which are being applied in forestry, agriculture, and horticulture. Specific procedures are required to multiply AM fungi and produce high quality inocula, and AM fungi are applied using different carriers, resulting in a wide range of formulations, including encapsulation. Recent developments in AM inoculum production systems include in vitro monoxenic root organ cultures.

3.2.2 AM and stress alleviation in plants

As indicated earlier, it is commonly accepted that AM associations helped plants to thrive in hostile environments during their evolution, and continue to help plants develop in stressed environments [18]. This is a critical AM role in the current scenario of climate change, where adverse conditions tend to exacerbate diverse stress situations which negatively affect the functionality/productivity of agricultural systems. In order to cope with adversity plants must develop adaptive strategies to increase their resilience and overcome negative impacts of natural or anthropogenic environmental stress conditions. AM symbiosis is considered an adaptive strategy since it increases plant tolerance of environmental constraints [17]. As a general phenomenon, plant perception of environmental stress cues triggers the activation of signaling molecules, (for which phytohormones are protagonists), which, once the signal is processed, ends with signal output enabling plants to respond to environmental constraints. Understanding how phytohormones act and interact in the signaling network, where jasmonic acid (JA) plays a key role, is fundamental to learning how AM plant systems thrive and survive in stressed environments. This understanding is essential to decipher the complex regulation of plant growth and immunity, and to design biotechnological strategies to optimize plant adaptation mechanisms and improve the ability of AM fungi and other soil microbes to alleviate stress in crops [34].

Apart from this hormone-controlled signaling cascade, other ecological, physiological, and molecular mechanisms are involved in the AM role in stress alleviation. These mechanisms have been the subject of diverse experimental studies and reviews over the last decade [18]. This information is briefly summarized in this chapter. The mechanisms involved in AM activity to increase resistance to/tolerance of pathogens and pests will be analyzed in depth in Section 3.3. We concentrate here on the mechanisms related to the alleviation of abiotic stressors (nutrient deficits, salinity, drought, and contamination).

The key AM fungi mechanisms for alleviation of the negative effects of nutrient deficits are based primarily on those described earlier: both the characteristics of the external mycelium, and the fact that gene expression related to the transport proteins is up-regulated under P and N deficit [28].

To alleviate the negative effects of osmotic stressors (drought, salinity...), plants must develop a number of adaptation mechanisms, mainly including fine regulation of their water uptake capacity and transpiration rate, and activation of the antioxidant machinery to overcome the overproduction of reactive oxygen species (ROS) caused by stress. These two general mechanisms (maintaining water and ROS balance) may be ameliorated by the establishment of AM symbiosis which acts via diverse specific mechanisms. These can be summarized as follows: (1) cell osmoregulation; (2) ionic homeostasis; (3) regulation of root water uptake and redistribution along plant tissues by aquaporins; (4) defense against antioxidants; and (5) maintenance of photosynthetic capacity. Such microbial activities result in better regulation of plant water status and contribute to increasing plant resistance to osmotic stress conditions. Finally, the improved water uptake capacity of microbized plants allows them to have higher transpiration rates and hence higher photosynthetic rates under conditions of water deficit. Particular attention is being paid to the role played by AM fungi in improving plant water status based on the improvement of root hydraulic conductance, which ultimately depends on aquaporin function. The available information on the subject was recently discussed [35–39].

3.2.3 Effects of agricultural practices on AM symbiosis

Agricultural practices such as tillage, fertilizer and pesticide application, monocultures, crop rotations with non-host plants, and fallow periods are known to have a significant negative effect on the natural biodiversity, abundance, and mycelial development of AM fungi in soils and on the establishment of AM symbiosis [17]. In uncultivated soils, host plants are colonized by extraradical hyphae, which constitute the main source of AM fungal inoculum [40]. In agricultural systems, soil tillage breaks up the network of extraradical AM hyphae formed by indigenous AM fungal populations, hindering colonization of the roots of cultivated plants and thereby reducing or preventing the benefits associated with symbiosis [41, 42]. The richness and abundance of AM fungal species and their spores have also been found to decrease in conventionally tilled soils relative to untilled soils [43, 44]. Similar effects on AM root colonization and spore abundance have been observed in agricultural systems of continuous monoculture, after non-host corps, or after long fallow periods [45–47].

Phosphorus (P) fertilization leading to a P supply in the soil which exceeds the crop's requirements frequently preclude mycorrhizal development, whilst low-input crop production systems may enhance mycorrhizal activity [48]. Under high P availability in soil, the P concentration in plant tissue increases to a level which causes a reduction both in the extent of root AM colonization and in the length of extraradical hyphae [49].

The reported effects of pesticides on AM symbiosis have been very varied. Recent studies [50, 51] have shown that glyphosate, a herbicide used globally, significantly decreases root mycorrhization and AM fungal spore viability and biomass in soil. However, no negative effects of nicosulfuron (used for the post-emergence control of weeds in maize) on mycorrhizal colonization and AM fungal richness in the field at dose rates ×1, ×2 and ×5 the recommended dose were found [52]. Increasing doses of fenpropimorph and fenhexamid, fungicides belonging to the family of sterol biosynthesis inhibitors (SBI), widely used in agriculture, have negative effects on hyphal development, physiology, and metabolism of AM fungi [53, 54]. In contrast, several insecticides (aldicarb, fenamiphos and dimethoate) did not affect AM symbiosis [55].

3.3 Interactions among plants, AM symbiosis, and mycotoxigenic fungi related to plant health

Before properly discussing the scarce information on the interactions between AM and mycotoxigenic fungi related crop plant health, we present a brief analysis of the role of AM fungi on plant protection against pathogens and pests and the mechanisms underlying such effects. Special attention will be given to the modulation of the plant immune system by AM fungi for induced systemic resistance and to priming for enhanced defense by AM fungi. These mechanisms may be involved in possible interactions between AM and mycotoxigenic fungi.

3.3.1 The effect of AM on plant protection against pathogens and pests

Diverse studies have shown the protective effect of AM colonization against infections by microbial pathogens, pests, and parasitic plants [19, 56, 57]. Remarkably, AM formation protects plants not only against soil-borne plant pathogens, but also against some which attack shoots. For the most part because of this systemic protection, the term mycorrhiza-induced resistance (MIR) was therefore coined [58].

In fact, AM symbiosis is known to reduce the damage caused by diverse soil-borne pathogens including fungi (*Fusarium, Rhizoctonia, Verticillium, Phytophthora,* and *Pythium*), bacteria, and parasitic nematodes [57]. Several mechanisms can operate simultaneously including competition with soil-borne pathogens for space and nutrients, altering root morphology or the quality and quantity of root exudates. The use of experimental split-root systems allowed us to confirm that protection by AM fungi is manifested in non-colonized areas of the root system. These experiments based on physical separation of AM fungi and the aggressor demonstrated the involvement of plant-mediated responses in enhancing plant resistance, as discussed later [57].

Furthermore, the ability of AM symbiosis to control shoot pathogens by eliciting a systemic, plant-mediated, resistance response was also shown [57, 58]. Transcriptional regulation of defense-related genes and accumulation of insect antifeedant compounds have been reported in the shoots of AM plants. The effect of AM symbiosis on controlling pathogens with a hemibiotrophic lifestyle varies from no effect to a reduction of the disease, but there is evidence of a more generalized positive effect of AM symbiosis in promoting plant resistance against necrotrophic shoot pathogens [57].

Multitrophic AM-plant-herbivore interactions have been the subject of diverse studies [57]. According to this review, AM symbiosis can influence the performance of insect herbivores, but the end result will depend on a compromise between a positive effect derived from enhanced plant growth and a negative effect derived from induced resistance in the plant, and depends on the type of insect attacking. Generalist insects are usually affected negatively by the presence of AM fungi, while specialist insects usually perform better on AM plants, probably because of the improved nutritional quality of the host. The degree of protection also depends on feeding type of the attacking insect. Phloem-sucking insects are little affected by the host immune response, while leaf chewers and miners are usually negatively affected by AM fungi.

Some plants (*Striga*, *Orobanche*, and *Phelipanche*) can parasitize a number of important crop plants and are considered amongst the most damaging agricultural pests [59]. López-Râez et al. [59] revised several experimental studies and concluded that germination and emergence of *Striga* seeds are reduced in AM plants. They discussed the potential role of strigolactones, a new class of plant hormones, emerging as a new biological control strategy against weeds. Strigolactones are exuded into soil, where they are known to act both as host detection signals for AM formation and as germination stimulants for root parasitic plant seeds. As strigolactone production is significantly reduced on establishment of AM symbiosis, a lowering of the germination rate of weed seeds is shown. This reduction therefore seems to be the mechanism underlying the decrease in incidence and damage of root parasitic weeds on crop plants when properly mycorrhizal and suggests the potential of AM symbiosis in controlling root parasitic weeds.

3.3.2 Mycorrhiza-induced resistance and priming of plant defenses

Both AM and pathogenic fungi have to cope with the plant immune system and diverse elicitors (microbe-associated molecular partners [MAMPs]) and effectors, the microbial molecules responsible for triggering resistance mechanisms, have been characterized. In AM fungi, lipochitooligosaccharides have been described as fundamental elicitors which activate responses in the host plants [30]. During the early stages of interaction in AM formation, the plant reacts to the presence of AM fungi by activating defense-related responses which are subsequently suppressed. Therefore, AM fungi have to modulate plant defense responses to achieve functional root colonization. Consequently, a mild but effective activation of the plant immune response occurs both locally and systemically [57].

Once the AM fungus is established in root tissue, the plant has to regulate the level of AM fungal proliferation within the roots, prevent colonization of vascular tissue, and avoid carbon drainage due to excessive colonization. Maintaining the equilibrium in this mutual symbiosis is associated with the modulation of the host plant's immune system by AM fungi, resulting in plant defense mechanisms not only regulating and controlling the symbiosis, but also directly impacting on root pathogens [19]. Such modulation may precondition plant tissue, allowing more efficient activation of defense mechanisms in response to attack by potential enemies, a phenomenon known as priming [58].

The mechanisms underlying the modulation of plant defense responses by AM fungi, which behave as a biotrophic microorganism, involve a quick but transient increase of endogenous salicylic acid (SA) in the roots combined with an accumulation of defensive compounds such as reactive oxygen species, specific isoforms of hydrolytic enzymes, and the activation of the phenylpropanoid pathway [40]. Recent studies demonstrate that AM fungi suppress plant defense reactions by secreting effector proteins which interfere with the host's immune system [60]. In the case of AM symbiosis, the levels of SA and other defense-related phytohormones such as jasmonic acid (JA), abscisic acid (ABA), and ethylene (ET) appear altered in AM roots, where ABA is also fundamental for AM function [61]. The regulation of defense signaling molecules involved in the modulation of plant immunity by AM fungi may play a major role in MIR which is also expressed in aboveground plant tissues, producing a systemic response effective against plant pathogens and herbivores. Interestingly, AM plants are more resistant to necrotrophs and chewing insects, which are sensitive to JA-dependent defense responses, while AM plants are more susceptible to biotrophs, which are sensitive to SA-regulated defenses. In fact, the activation of JA-dependent and repression of SA-dependent defenses have been demonstrated during development of AM symbiosis [58, 59, 61, 62].

A key point to consider is that the constitutive expression of defenses is too costly for the plant in the absence of challenging attackers. However, some microorganisms have developed the ability to enhance resistance not only via direct activation of defenses but also by inducing sensitization of the tissues on appropriate stimulation to express basal defense mechanisms more efficiently after a subsequent pathogen attack. This preconditioning of plant tissues is known as the priming of the plant defense [58, 59, 63]. Priming maintains the plant in an "alert" state in which defenses are not actively expressed but get the plant prepared to respond to any attack in a faster and/or stronger manner in comparison to plants which have not had the priming stimulus. This results in increased plant resistance. Priming therefore confers important plant fitness benefits in terms of plant health [57]. Additionally, preconditioning of plant tissues for a quick and more effective activation of defenses on attack has important ecological fitness benefits compared to direct activation of defenses. This boost (priming) of basal defenses is successfully developed by AM fungi and can be considered the main mechanism operating in MIR, as deduced by the stronger defense reactions triggered in the AM plant upon attack [58].

Studies on primed defense responses in AM plants have been reviewed [58]. Preliminary studies referred to primed roots and include experiments involving attack by *Fusarium* (AM transformed carrot roots), *Phytophthora parasitica* (tomato), *Rhizoctonia* (potato), and *F. oxysporum* (date palm trees). It was shown that the primed response is not restricted to the root system but also operates in the shoots of AM plants. In fact, in tomato plants AM symbiosis induced systemic resistance against the necrotrophic foliar pathogen *Botrytis cinerea*. It was found that the amount of pathogen in the leaves of AM plants was significantly lower, while the expression of some JA-regulated, defense-related genes was higher in AM plants [64]. Gene expression and enzymatic activities were monitored in a time-course experiment after shoot (tomato) treatment with JA, ET, and SA. Transcript profiling of the leaves of AM and non-AM plants on treatment with JA indicated a stronger induction of JA-regulated genes in AM plants, including typical defense-related JA responsive genes supporting a prominent role of priming for JA-dependent responses in AM-induced resistance [19].

3.3.3 Interactions between AM symbiosis and mycotoxigenic fungi

It would be expected that the prophylactic ability of AM symbiosis could be applied to the biological control of mycotoxigenic fungi. However, to our knowledge the published information on this topic is scarce. Using confrontation cultures Ismail et al. [65] reported that the growth of *Fusarium sambucinum*, the causal agent of tuber sprout rotting in potato [66], was significantly reduced in the presence of the AM fungus *Glomus irregulare*, and that the relative expression of the *TRI5* gene in *F. sambucinum* was up-regulated and that of *TR101* gene greatly down-regulated after confrontation with isolates of *G. irregulare*. The predominant mycotoxin of *F. sambucinum* is 4,15-diacetoxyscirpenol (DAS), a trichothecene toxic to humans and animals [65, 67]. *TR15* encodes the first enzyme involved in the trichothecene biosynthesis pathway and *TR101* encodes a trichothecene 3-O-acetyltransferase which acetylates the C-3 of various *Fusarium* trichothecenes, converting them to less toxic products [68, 69].

The role of *G. irregulare* on the biocontrol of *F. sambucinum* on potato has been studied [70]. The authors demonstrated that inoculation of the AM fungus significantly reduced disease severity of *F. sambucinum* on mycorrhizal potato plants with respect to those infected and non-mycorrhizal, also decreasing the negative effects of the pathogen on biomass and potato tuber production. The study also showed that *G. irregulare* could regulate expression of certain defense-related potato genes and thus promote resistance against *F. sambucinum*.

Although further research is needed, the scarce information available suggests that AM symbiosis may play an important role in the protection of cultivated plants against pathogenic *Fusarium* species and their mycotoxins, through the regulation of defense-related genes and/or the modulation of mycotoxin gene expression.

3.3.4 Impact of climate change on AM fungi and repercussions for the protection of food crops against fungal diseases

Intensive agriculture production is fundamental to satisfy food requirements for the growing world population. However, its realization is associated with the mass consumption of non-renewable natural resources, with environmental contamination, and with the emission of greenhouse gases causing climate change. Society and science are becoming aware that sustainable agricultural approaches are therefore needed, but the impacts of climate change must be considered because of their influence on soil fertility, crop nutrition, and on the production of healthy food [1, 2]. Actually, evidence is accumulating to show that the CO_2 emissions derived from man-mediated activities are submitting the earth to warming regardless of the mitigation strategies now applied [20], therefore the mitigation approaches must be improved. Factors involved in climate change, such as elevated CO_2 levels, are known to influence the incidence and effect of plant diseases, particularly the appearance of new diseases, and also to influence the performance of mycotoxigenic fungi, as discussed in other chapters. We will therefore deal specifically with AM fungi.

How climate change affects the activity of plant-associated microorganisms related to nutrient cycling or to disease occurrence has been the subject of a number of analyses [20, 71, 72]. The general conclusion of these studies is that the information is fragmentary and that much more research is needed to understand and predict plant responses to global changes under natural conditions. Actually, our ability to predict the impact of global environmental change on plant-microbiome interactions and to know how the biotic responses to such changes are regulated is hampered because of our limited understanding of these interactions.

The first analyses on the impact of climate change on AM formation and function, reviewed ten years ago, mainly addressed natural ecosystems, and either studied the potential of AM symbiosis to mitigate the impact of global change on plant communities [73] or described the response of AM symbiosis to elevated atmospheric CO_2 [74]. It appears that the effects on AM fungi were largely controlled by host-plant responses, however, these fungi can respond directly to elevated soil temperature, a fact which may have large implications for the rates of C cycling. Evidence shows that AM fungal hyphae may have a very short life, resulting in rapid return of C to the atmosphere; evaluation of the impact of soil temperature on hyphal turnover is therefore fundamental [74].

More recent review articles support that predicted global surface temperature rise affects AM fungi. The general conclusion of one of these studies [24] is that soil temperature increase has a positive impact on AM fungal root colonization and hyphal length. Another review article [75] emphasizes that rising global atmospheric CO_2 levels also affect AM fungal diversity, abundance, and function. As an increase in CO_2 enhances photosynthetic rates, more C is directed to the soil pool and AM fungi, which will benefit AM services, including nutrient cycling and plant health promotion. In contrast, anthropogenic N deposition reduces AM fungal diversity and abundance; however literature describing the effects of elevated CO_2 and N deposition interaction on AM fungi is scarce.

In an exhaustive analysis of the rhizosphere responses to elevated CO_2 [22], it was found that an increased allocation of CO_2 from photosynthesis is directed to mycorrhizas, which then translocate C into microbial communities. This is traduced in changes in structure, size, and activities of the plant associated microbiome in the mycorrhizosphere. These changes result in an increased rate of respiration and the subsequent return of CO_2 to the atmosphere.

Environmental changes, such as elevated CO_2, are recognized as a key factor influencing the incidence of plant diseases or the appearance of new diseases [2, 7]. All in all, predictions have considerable uncertainty and future disease trends may be different for each geographic region [76].

3.3.5 Research perspectives and opportunities for exploiting the interactions between mycotoxigenic and AM fungi with regard to plant health as affected by climate change

Implementation of control practices for mycotoxigenic fungi is necessary to produce healthy foods. Because sustainable agricultural practices must be followed, the logical techniques are those based on applying biological control approaches. Here we propose the use of AM fungi for such purposes. In fact, AM fungi are considered a biotechnological tool for linking biotic and abiotic agricultural facts involved in healthy crop production in a climate change scenario [77].

Concerning the interactions between mycotoxigenic and AM fungi in relation to plant health, it is necessary to propose research approaches, firstly, to improve our understanding of these interactions, and secondly to establish appropriate conditions allowing the expression of the AM abilities to control the pathogen. To the author's knowledge, the first available studies in this field of research are those discussing the possibility of using AM fungi for the biocontrol of the mycotoxigenic fungus *Fusarium sambucinum* [65, 70]. However, in order to further develop the use of AM fungi for the biological control of mycotoxigenic fungi, other considerations related to the prophylactic possibilities of AM fungi must be taken into account. In this context, a key mechanism for efficient protection against pathogens is defense priming, the preconditioning of immunity induced by microbial colonization after seed germination [78]. AM fungi are particularly recognized for their ability to prime plant defense [19], and have coevolved with plants by establishing AM symbiosis, also evolving developmental signals involved in immunity maturation [28].

In the future, the exploitation of AM symbiosis to protect plants against deleterious organisms, particularly fungal pathogens, is highly recommended by exploiting the recently established fact that AM colonization can prime plant immunity by boosting plant defenses to respond to pathogen attack, as part of the "mycorrhiza-induced systemic resistance" ability of AM fungi [19]. This ability must be explored using AM fungi isolated from the rhizosphere of the target food crop plants to be screened in greenhouse assays challenging mycotoxigenic fungi. Such studies under controlled conditions may help improve our understanding of the interactions between AM and mycotoxigenic fungi and to define the conditions where mycotoxin production can be reduced. The selected AM fungi must then be tested in small-scale studies under field conditions. All in all, we need to improve our knowledge of the interactions among plants, mycotoxigenic and AM fungi, to establish biocontrol measurements in a climate change scenario. Meta-analyses aimed at analyzing the responses of plants and their fungal associates to global change factors are recommended in order to allow us to establish general trends in the responses of plant–fungal symbioses to future environments.

A model study on this subject [79] was based on a meta-analysis of 434 studies analyzing the responses of plants and their associated fungal symbionts, including class I leaf endophytes and AM fungi, to four global change factors (enriched CO_2, drought, N deposition, and warming). The class I leaf endophytes are known to protect plants against insect pests and vertebrate herbivores, drought, and nutrient deficits, but also to produce secondary metabolites toxic for animals. In general, fungal symbionts increase plant biomass and ameliorate the negative effects of global change, but the meta-analysis revealed gaps in current research analyzing the responses of plant–fungal symbioses to global change. The authors [79] confirm that it is critical to consider plant-fungal symbiosis to predict plant responses to global change. Most studies in the meta-analysis were conducted within the greenhouse and therefore may not reflect responses under field conditions. Multiple fungal symbionts were analyzed by means of paired studies aimed at analyzing effects on plant performance, but hardly commented on how mycosymbionts interact amongst themselves, and did not focus on biological control of AM fungi over mycotoxigenic fungi.

In conclusion, several issues can be pointed out. Probably the most critical point is that information on interactions between symbiotic fungi, such as mutualistic AM and mycotoxigenic fungi (pathogenic *Fusarium* and others in cereals), is scarce. Only a small number of publications refer to changes in the colonization rates of the symbionts involved, which is important in controlling pathogen development. Considering the prophylactic ability of AM fungi, more research on their interactions with mycotoxigenic fungi are highly recommended. This may be more urgently needed in the current/future scenario of climate change, where AM services appear fundamental for the production of healthy crops, and where the impact of mycotoxigenic fungi appears to pose a threat to food security. In fact, AM symbiosis can be considered an adaptive strategy which provides the plant with an increased ability to alleviate stress, be it biotic (pathogen attack) or abiotic (nutrient deficit, drought, salinity, etc.).

For the aims of this chapter we have concentrated on biotic stress alleviation. As stated earlier, it is fundamental to remember that AM symbiosis protects plants against deleterious organisms, including microbial pathogens, herbivorous insects, and parasitic plants. Furthermore, the ability of AM symbiosis to control shoot pathogens by eliciting a systemic, plant-mediated resistance response has also been shown. Results from the last few years have clearly shown that AM colonization can prime plant immunity by boosting the plant's ability to respond to pathogen attack, as part of the "mycorrhiza-induced systemic resistance". The mechanisms involved in modulation of the host plant's immune system by AM fungi and the priming of plant defense have been established. These findings support the scientific basis for proposing appropriate research approaches to understand how mycotoxigenic and AM fungi interact. This is relevant in order to define perspectives, challenges, and opportunities for exploiting AM services to control the activities of mycotoxigenic fungi, with regards to promoting food security in the current scenario of climate change.

References

[1] St Clair SB, Lynch JP. The opening of Pandora's Box: climate change impacts on soil fertility and crop nutrition in developing countries. *Plant Soil* 2010;335:101–115.

[2] Dwivedi S, Sahrawat K, Upadhyaya H, Ortiz R. Food, nutrition and agrobiodiversity under global climate change. *Adv Agron* 2013;120:1–128.

[3] Chakraborty S, Pangga IB, Roper MM. Climate change and multitrophic interactions in soil: the primacy of plants and functional domains. *Glob Change Biol* 2012;18:2111–2125.

[4] Coakley SM, Scherm H, Chakraborty S. Climate change and plant disease management. *Annu Rev Phytopathol* 1999;37:399–426.

[5] Roos J, Hopkins R, Kvarnheden A, Dixelius C. The impact of global warming on plant diseases and insect vectors in Sweden. *Eur J Plant Pathol* 2011;129:9–19.

[6] Pautasso M, Doering TF, Garbelotto M, Pellis L, Jeger MJ. Impacts of climate change on plant diseases-opinions and trends. *Eur J Plant Pathol* 2012;133:295–313.

[7] West JS, Townsend JA, Stevens M, Fitt BDL. Comparative biology of different plant pathogens to estimate effects of climate change on crop diseases in Europe. *Eur J Plant Pathol* 2012;133:315–331.

[8] Newton AC, Johnson SN, Gregory PJ. Implications of climate change for diseases, crop yields and food security. *Euphytica* 2011;179:3–18.

[9] FAO 2009. World Summit on Food Security 2009. *Declaration of the World Summit on Food Security. WSFS 2009/2.* Rome: Food and Agriculture Organization of the United Nations (FAO). Available at http://www.fao.org/fileadmin/templates/wsfs/Summit/Docs/Final_Declaration/WSFS09_Declaration.pdf.

[10] FAO. 2014. *Crop Prospects and Food Situation № 2.* Rome: Trade and Markets Division, Food and Agriculture Organization of the United Nations (FAO). Available at http://www.fao.org/3/a-i3899e.pdf.

[11] Alexandratos N, Bruinsma J. *World agriculture towards 2030/2050: the 2012 revision.* Rome: Food & Agriculture Organization of the United Nations Statistics. 2012.

[12] Chakraborty BN, Chakraborty U, Rai K, Allay S, Dey PL. Molecular detection of fungal pathogens of *Citrus reticulata* grown in Darjeeling hills and their management. II International Symposium on Citrus Biotechnology 2011;892:189–206.

[13] Paterson RRM, Lima N. How will climate change affect mycotoxins in food? *Food Res Int* 2010;43:1902–1914.

[14] Magan N, Medina A, Aldred D. Possible climate-change effects on mycotoxin contamination of food crops pre- and postharvest. *Plant Pathol* 2011;6:150–163.

[15] Orrell P, Bennett AE. How can we exploit above–belowground interactions to assist in addressing the challenges of food security? *Front Plant Sci* 2013;4.

[16] Smith SE, Read DJ. *Mycorrhizal Symbiosis*. 3rd ed. New York: Elsevier Academic Press; 2008.

[17] Jeffries P, Barea JM. Arbuscular Mycorrhiza – a key component of sustainable plant-soil ecosystems. In: Hock B (ed). *The Mycota, a comprehensive treatise on fungi as experimental systems for basic and applied research*. Berlin, Heidelberg: Springer; 2012. p. 51–75.

[18] Barea JM, Pozo MJ, López-Ráez JA, Aroca R, Ruíz-Lozano JM, Ferrol N, Azcón R, Azcón-Aguilar C. Arbuscular Mycorrhizas and their significance in promoting soil-plant systems sustainability against environmental stresses In: Rodelas B, González-López J. (eds). *Beneficial Plant-Microbial Interactions: Ecology and Applications*. Boca Raton, USA: CRC Press; 2013. p. 353–387.

[19] Pozo MJ, Jung SC, Martínez-Medina A, López-Ráez JA, Azcón-Aguilar C, Barea JM. Root allies: Arbuscular mycorrhizal fungi help plants to cope with biotic stresses. In: Aroca R (ed.) *Symbiotic Endophytes*. Berlin Heidelberg: Springer-Verlag; 2013. p. 289–307.

[20] Compant S, van der Heijden M, Sessitsch A. Soil warming effects on beneficial plant-microbe interactions. In: de Bruijn FJ (ed). *Molecular Microbial Ecology of the Rhizosphere*. Hoboken, New Jersey, USA: John Wiley & Sons; 2013. p. 1045–1054.

[21] Rodríguez AA, Córdoba AR, Ortega L, Taleisnik E. Decreased reactive oxygen species concentration in the elongation zone contributes to the reduction in maize leaf growth under salinity. *J Exp Bot* 2004;55:1383–1390.

[22] Drigo B, Kowalchuk GA. Rhizosphere responses to elevated CO_2. In: de Bruijn FJ (ed.) *Molecular Microbial Ecology of the Rhizosphere*. Hoboken, New Jersey, USA: John Wiley & Sons; 2013. p. 1063–1074.

[23] Schüßler A, Walker C. Evolution of the 'plant-symbiotic' fungal phylum, Glomeromycota. In: Pöggeler S, Wöstemeyer J (eds). *Evolution of Fungi and Fungal-Like Organisms*. Berlin, Heidelberg: Springer-Verlag; 2011. p. 163–185.

[24] Zuccaro A, Lahrmann U, Langen G. Broad compatibility in fungal root symbioses. *Curr Opin Plant Biol* 2014;20:135–145.

[25] Porras-Alfaro A, Bayman P. Hidden fungi, emergent properties: Endophytes and microbiomes. *Ann Rev Phytopathol* 2011;49:291–315.

[26] Mercado-Blanco, J. 2015. Life of microbes inside the plant. In: Lugtenberg B (ed). *Principles of Plant-Microbe Interactions*. Switzerland, Heidelberg: Springer International Publishing; 2015. p. 25–32.

[27] Malfanova N, Lugtenberg BJJ, Berg G. Bacterial endophytes: who and where, and what are they doing there? In: de Bruijn FJ (ed). *Molecular Microbial Ecology of the Rhizosphere*. Hoboken, New Jersey, USA: Wiley Blackwell; 2013. p. 393–403.

[28] Barea JM, Azcón-Aguilar C. Evolution, biology and ecological effects of arbuscular mycorrhiza. In: Camisao AF, Pedroso CC (eds). *Symbiosis: Evolution, Biology and Ecological Effects*. New York: Nova Science; 2013. p. 1–34.

[29] Schüßler A, Gehrig H, Schwarzott D, Walker C. Analysis of partial *Glomales* SSU rRNA gene sequences: implications for primer design and phylogeny. *Mycol Res* 2001;105:5–15.

[30] Corradi N, Bonfante P. The arbuscular mycorrhizal symbiosis: Origin and evolution of a beneficial plant infection. *PLoS Pathog* 2012;8(4);e1002600. Doi: 10.1371/jounal.ppat.1002600.

[31] Ferrol N, Pérez-Tienda J. Coordinated nutrient exchange in arbuscular mycorrhiza. In: Azcón-Aguilar C, Barea JM, Gianinazzi S, Gianinazzi-Pearson V (eds). *Mycorrhizas Functional Processes and Ecological Impact*. Berlin, Heidelberg: Springer-Verlag; 2009. p. 73–87.

[32] Recorbet G, Abdallah C, Renaut J, Wipf D, Dumas-Gaudot E. Protein actors sustaining arbuscular mycorrhizal symbiosis: underground artists break the silence. *New Phytol* 2013;199:26–40.

[33] Fester T, Sawers R. Progress and challenges in agricultural applications of arbuscular mycorrhizal fungi. *Crit Rev Plant Sci* 2011;30:459–470.

[34] Pozo M, López-Ráez J, Azcón-Aguilar C, García-Garrido JM. Phytohormones as integrators of environmental signals in the regulation of mycorrhizal symbioses. *New Phytol* 2015;205:1431–1436.

[35] Ruíz-Lozano JM, Porcel R, Azcón C, Aroca R. Regulation by arbuscular mycorrhizae of the integrated physiological response to salinity in plants: new challenges in physiological and molecular studies. *J Exp Bot* 2012;63:4033–4044.

[36] Ruíz-Lozano JM, Porcel R, Barzana G, Azcón R, Aroca R. Contribution of arbuscular mycorrhizal symbiosis to plant drought tolerance: state of the art. In: Aroca R (ed). *Plant Responses to Drought Stress From Morphological to Molecular Features*. Heidelberg, Germany: Springer; 2012. p. 335–362.

[37] Aroca R, Porcel R, Ruíz-Lozano JM. Regulation of root water uptake under abiotic stress conditions. *J Exp Bot* 2012;63:43–57.

[38] Porcel R, Aroca R, Ruíz-Lozano JM. Salinity stress alleviation using arbuscular mycorrhizal fungi. A review. *Agron Sustain Develop* 2012;32:181–200.

[39] Calvo-Polanco M, Sánchez-Romera B, Aroca R. Arbuscular mycorrhizal fungi and the tolerance of plants to drought and salinity. In: Aroca R (ed). *Symbiotic Endophytes*. Berlin, Heidelberg: Springer-Verlag; 2013. p. 271–288.

[40] Bárzana G, Aroca R, Bienert GP, Chaumont F, Manuel Ruíz-Lozano J. New Insights into the regulation of aquaporins by the arbuscular mycorrhizal symbiosis in maize plants under drought stress and possible implications for plant performance. *Mol Plant-Microbe Interact* 2014;27:349–363.

[41] McGonigle TP, Miller MH. Development of fungi below ground in association with plants growing in disturbed and undisturbed soils. *Soil Biol Biochem* 1996;28:263–269.

[42] Kabir Z. Tillage or no-tillage: Impact on mycorrhizae. *Can J Plant Sci* 2005;85:23–29.

[43] Boddington CL, Dodd JC. The effect of agricultural practices on the development of indigenous arbuscular mycorrhizal fungi. I. Field studies in an Indonesian ultisol. *Plant Soil* 2000;218:137–144.

[44] Curaqueo, G, Barea, JM, Acevedo, E, Rubio, R, Cornejo, P, Borie, F. Effects of different tillage system on arbuscular mycorrhizal fungal propagules and physical properties in a Mediterranean agroecosystem in central Chile. *Soil Till Res* 2011;113:11–18.

[45] Smith TF. The effect of season and crop-rotation on the abundance of spores of vesicular-arbuscular (VA) mycorrhizal endophytes. *Plant Soil* 1980;57:475–479.

[46] Sieverding E, Leihner DE. Influence of crop-rotation and intercropping of cassava with legumes on va mycorrhizal symbiosis of cassava. *Plant Soil* 1984;80:143–146.

[47] Thompson JP. Decline of vesicular-arbuscular mycorrhizae in long fallow disorder of field crops and its expression in phosphorus deficiency of sunflower. *Aust J Agric Res* 1987;38:847–867.

[48] Grant C, Bittman S, Montreal M, Plenchette C, Morel C. Soil and fertilizer phosphorus: Effects on plant P supply and mycorrhizal development. *Can J Plant Sci* 2005;85:3–14.

[49] Liu A, Hamel C, Hamilton RI, Ma BL, Smith DL. Acquisition of Cu, Zn, Mn and Fe by mycorrhizal maize (*Zea mays* L.) grown in soil at different P and micronutrient levels. *Mycorrhiza* 2000;9:331–336.

[50] Druille M, Cabello MN, Omacini M, Golluscio RA. Glyphosate reduces spore viability and root colonization of arbuscular mycorrhizal fungi. *Appl Soil Ecol* 2013;64:99–103.

[51] Zaller JG, Heigl F, Ruess L, Grabmaier A. Glyphosate herbicide affects belowground interactions between earthworms and symbiotic mycorrhizal fungi in a model ecosystem. *Scientific Reports* 2014;4.

[52] Karpouzas DG, Papadopoulou E, Ipsilantis I, Friedel I, Petric I, Udikovic-Kolic N, Djuric S, Kandeler E, Menkissoglu-Spiroudi U, Martin-Laurent F. Effects of nicosulfuron on the abundance and diversity of arbuscular mycorrhizal fungi used as indicators of pesticide soil microbial toxicity. *Ecol Indicators* 2014;39:44–53.

[53] Zocco D, Fontaine J, Lozanova E, Renard L, Bivort C, Durand R, Grandmougin-Ferjani A, Declerck S. Effects of two sterol biosynthesis inhibitor fungicides (fenpropimorph and fenhexamid) on the development of an arbuscular mycorrhizal fungus. *Mycol Res* 2008;112:592–601.

[54] Zocco D, Van Aarle IM, Oger E, Lanfranco L, Declerck S. Fenpropimorph and fenhexamid impact phosphorus translocation by arbuscular mycorrhizal fungi. *Mycorrhiza* 2011;21:363–374.

[55] Schweiger PF, Jakobsen I. Dose-response relationships between four pesticides and phosphorus uptake by hyphae of arbuscular mycorrhizas. *Soil Biol Biochem* 1998;30:1415–1422.

[56] Lugtenberg BJJ, Malfanova N, Kamilova F, Berg G. Microbial control of plant root diseases. In: de Bruijn FJ (ed). *Molecular Microbial Ecology of the Rhizosphere*. Hoboken, New Jersey, USA: Wiley Blackwell; 2013. p. 575–586.

[57] Jung SC, Martínez-Medina A, López-Ráez JA, Pozo MJ. Mycorrhiza-induced resistance and priming of plant defenses. *J Chem Ecol* 2012;38:651–664.

[58] Pozo MJ, Azcón-Aguilar C. Unraveling mycorrhiza-induced resistance. *Curr Opin Plant Biol* 2007;10:393–398.

[59] López-Ráez JA, Bouwmeester H, Pozo MJ. Communication in the rhizosphere, a target for pest management In: Lichtfouse E (ed). *Sustainable Agriculture Reviews vol. 8 Agroecology and Strategies for Climate Change*. Netherlands: Springer; 2012. p. 109–133.

[60] Kloppholz S, Kuhn H, Requena N. A secreted fungal effector of *Glomus intraradices* promotes symbiotic biotrophy. *Curr Biol* 2011;21:1204–1209.

[61] López-Ráez JA, Pozo MJ, García-Garrido JM. Strigolactones: a cry for help in the rhizosphere. *Botany-Botanique* 2011;89:513–522.

[62] López-Ráez JA, Charnikhova T, Fernández I, Bouwmeester H, Pozo MJ. Arbuscular mycorrhizal symbiosis decreases strigolactone production in tomato. *J Plant Physiol* 2011;168:294–297.

[63] Pozo MJ, Verhage A, García-Andrade J, García JM, Azcón-Aguilar C. Priming plant defence against pathogens by arbuscular mycorrhizal fungi. In: Azcón-Aguilar C, Barea JM, Gianinazzi S, Gianinazzi-Pearson V (eds). *Mycorrhizas Functional Processes and Ecological Impact*. Berlin, Heidelberg: Springer-Verlag; 2009. p. 123–135.

[64] Pozo MJ, Jung SC, López-Ráez JA, Azcón-Aguilar C. Impact of arbuscular mycorrhizal symbiosis on plant response to biotec stress: The role of plant defence mechanisms. In: Koltai H, Kapulnik Y (eds). *Arbuscular Mycorrhizas: Physiology and Function*. Netherlands: Springer; 2010. p. 193–207.

[65] Ismail Y, McCormick S, Hijri M. A fungal symbiont of plant-roots modulates mycotoxin gene expression in the pathogen *Fusarium sambucinum*. PloS One 2011;6(3):E17990. doi:10.1371/journal.pone.0017990.

[66] Wharton PS, Tumbalam P, Kirk WW. First report of potato tuber sprout rot caused by *Fusarium sambucinum* in Michigan. *Plant Dis* 2006;90:1460.

[67] Desjardins AE, Hohn TM. Mycotoxins in plant pathogenesis. *Mol Plant-Microbe Interact* 1997;10:147–152.

[68] McCormick SP, Alexander NJ, Trapp SE, Hohn TM. Disruption of TRI101, the gene encoding trichothecene 3-O-acetyltransferase, from *Fusarium sporotrichioides*. *Appl Environ Microbiol* 1999;65:5252–5256.
[69] Garvey GS, McCormick SP, Rayment I. Structural and functional characterization of the TRI101 trichothecene 3-O-acetyltransferase from *Fusarium sporotrichioides* and *Fusarium graminearum* - Kinetic insights to combating fusarium head blight. *J Biol Chem* 2008;283:1660–1669.
[70] Ismail Y, Hijri M. Arbuscular mycorrhisation with *Glomus irregulare* induces expression of potato PR homologues genes in response to infection by *Fusarium sambucinum*. *Funct Plant Biol* 2012;39:236–245.
[71] Lynch JP, St Clair SB. Mineral stress: the missing link in understanding how global climate change will affect plants in real world soils. *Field Crops Res* 2004;90:101–115.
[72] Pritchard SG. Soil organisms and global climate change. *Plant Pathol* 2011;60:82–99.
[73] Rodríguez RJ, Redman RS, Henson JM. The role of fungal symbioses in the adaptation of plants to high stress environments. *Mitig Adapt Strat Gl* 2004;9:261–272.
[74] Fitter AH, Heinemeyer A, Husband R, Olsen E, Ridgway KP, Staddon PL. Global environmental change and the biology of arbuscular mycorrhizas: gaps and challenges. *Can J Bot-Rev Can Bot* 2004;82:1133–1139.
[75] Willis A, Rodrigues BF, Harris PJC. The ecology of arbuscular mycorrhizal fungi. *Crit Rev Plant Sci* 2013;32:1–20.
[76] Juroszek P, von Tiedemann A. Climate change and potential future risks through wheat diseases: a review. *Eur J Plant Pathol* 2013;136:21–33.
[77] Polle A, Luo ZB. Biotic and abiotic interactions in plants: Novel ideas for agriculture and forestry in a changing environment. Environ Exp Bot 2014;108:1–3.
[78] Selosse MA, Bessis A, Pozo MJ. Microbial priming of plant and animal immunity: symbionts as developmental signals. *Trends Microbiol* 2014;22:607–613.
[79] Kivlin SN, Emery SM, Rudgers JA. Fungal symbionts alter plant responses to global change. *Am J Bot* 2013;100:1445–1457.

Angel Medina, Alicia Rodriguez, and Naresh Magan

4 Changes in environmental factors driven by climate change: effects on the ecophysiology of mycotoxigenic fungi

4.1 Background

4.1.1 Environmental change, fungal adaptation, and mycotoxins

The famous philosopher Jean-Jacques Rousseau wrote, back in 1778, a sentence which is still current when applied to the knowledge in contemporary ecology: "Everything is in constant flux on this earth" [1].

Living organisms live in an ever-changing world and are exposed to constant environmental fluctuations in their surroundings (biotic, e.g. other species; and abiotic, e.g. temperature, water availability, gas composition, substrate). This requires them to adapt relatively quickly for effective exploitation of resources and to compete effectively in a particular ecosystem.

Filamentous fungi are a group of eukaryotes with tremendous physiological plasticity which has contributed to their capacity for adaptation and facilitated their active colonization in a range of extreme environments [2, 3]. They can be competitive in such environments because of their ability to produce a battery of extracellular hydrolytic enzymes, secondary metabolites, and volatiles (Fig. 4.1).

A number of *Aspergillus* and *Penicillium* species are xerotolerant or xerophilic and able to colonize intermediate moisture food products (reduced water activity; a_w) which represent environmentally stressed conditions for many other fungi and bacteria (Tab. 4.1). This physiological plasticity gives these species an important ecological advantage which allows them to successfully colonize a range of different food matrices.

As a result, many staple crops (cereals, nuts, fruits) can be colonized and infected by *Aspergillus/Penicillium* and *Fusarium* species which can then contaminate the edible parts with toxic secondary metabolites, the so-called mycotoxins. Although *Fusarium* species are not xerophiles, they can often colonize crops preharvest and produce mycotoxins when they are under relative water and temperature stress (0.90–0.93 a_w; 10–15, > 30 °C) for these species.

Mycotoxins have significant toxicological impact on both human and animal health. This has resulted in strict legislative limits for mycotoxins in many parts of the world in a wide range of foodstuffs with the strictest limits in the EU [4]. However, in many African countries where legislation is often applied to export crops only, consumption of mycotoxin contaminated staple foods poses a significant risk, espe-

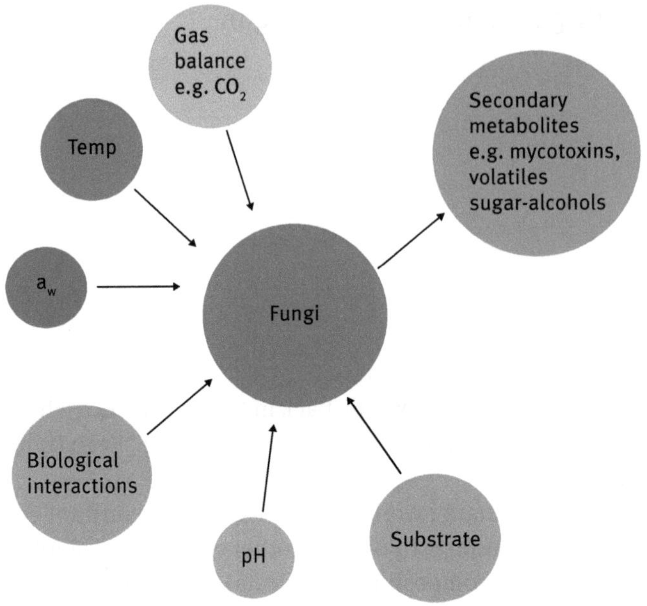

Fig. 4.1: Illustration showing the effect of different biotic and abiotic factors on the ecophysiology of fungi.

cially in rural populations and subgroups such as children and immunocompromised people.

The most important mycotoxins are aflatoxins (produced by *Aspergillus* section *Flavi* species), ochratoxin A (OTA) (*Aspergillus* section *Circumdati* species, *Aspergillus* section *Nigri* species, *Penicillium nordicum, P. verrucosum*), fumonisins (*Fusarium verticillioides, F. sporotrichioides*), type A trichothecenes, T-2 and HT-2 toxin (*F. langsethiae, F. sporotrichioides*), type B trichothecenes (*F. graminearum* and related species), and patulin (*P. expansum*). The European Food Safety Authority (EFSA) is now examining the relative hazard posed by mycotoxins produced by *Alternaria* species across the EU to obtain information of the levels in different food products and make decisions as to whether limits should also be considered.

4.1.2 Climate change and mycotoxigenic fungi

Agricultural production of staple foodstuffs requires optimized agronomic practices to provide the yields and amount of raw food commodities required to feed an ever-increasing global population. Thus, climate change scenarios could have a profound impact on the production and delivery of sufficient food. Climate change is expected to have a great effect on our landscape worldwide and the way we interact with it. Climate change models have projected a marked decrease in summer precipitation

Tab. 4.1: Food products, their moisture content and equivalent water activity, and the predominant microbial genera which can colonize and cause spoilage.

Type of product	**Water content (% w/w)**	**Water activity (a_w)**	**Common spoilage microorganisms**
Fresh meat and fish		0.99	Bacteria, yeasts
Carbonated drinks		0.999–0.977	Bacteria, yeasts
Cream, 25 % fat		0.973	Bacteria
Bread		0.95	Yeasts and molds
Muffins		≈ 0.95	Yeasts and molds
Salami, dry		0.875	Molds
Sausage, snack, fully dry	29–10.6	0.87–0.66	Molds
Aged cheddar		0.85	Yeasts and molds
Jams and jellies		0.80	Yeasts and molds
Plum pudding		0.80	Yeasts and molds
Fruits, dried		0.80–0.72	Xerophilic molds
		0.78	*Aspergillus flavus*
		0.77	*Aspergillus niger*
Wheat	≤ 14.5	≤ 0.70	GRAS
	> 14.5	> 0.70	Xerophilic molds
Maize	≤ 14	≤ 0.70	GRAS
	> 14	> 0.70	Xerophilic molds
Rice	≤ 12–15	≤ 0.70	GRAS
	> 12–15	> 0.70	Xerophilic molds
		0.70	*Eurotium amstelodami*
		0.656	*Xeromyces bisphorus*
Biscuits		0.30	GRAS
Honey		0.552	GRAS
Milk powder		0.20	GRAS
Instant coffee		0.20	GRAS

GRAS: Generally recognized as safe

and increases in CO_2 (double or triple existing concentrations) and temperature for some areas, which would result in concomitant drought stress episodes.

The EU green paper on climate change in Europe has suggested that effects will be regional and either be detrimental or advantageous depending on geographical area [5]. Even if some changes in climate may be positive for some northern European regions, many will be negative, affecting regions already suffering from environmental or other changes. According to the intergovernmental panel on climate change [6], the worst consequences may not be felt until 2050, but significant adverse effects are expected even in the short term from more frequent extreme conditions.

Generally, the atmospheric concentration of CO_2 is expected to at least double (from 350–400 to 800–1000 ppb). Because of this increase and that of other greenhouse gases, the global temperature is expected to increase by between +2 and +5 °C depending on levels of industrial activity. Hence, changes in southern Europe may equate to an increase of 4–5 °C with important reductions in rainfall and longer drought periods, resulting in increasing desertification, and a decrease in crop yields and arable areas. In areas of western and Atlantic Europe, changes of 2.5–3.5 °C with dryer and hotter summers are envisaged. In Central Europe, an increase of 3–4 °C, higher rainfall and floods are forecast, although longer growing periods may benefit crop yields. Northern Europe would expect a mean temperature increase of 3–4.5 °C, with a significant increase in precipitation of 30–40 %. This may lead to increases in crop yields and perhaps new crop cultivation patterns [5, 7]. A recent EFSA report [8] suggests that because of early ripening in cereals, especially wheat and maize, exposure to plant diseases and possibly increased contamination with aflatoxins can be expected in the next 25 years in most of southern and central Europe. Similar effects have been described in other areas of the world, especially parts of Asia and Central and South America, which are important producers of staple crops [6].

Such environmental changes are slowly but steadily shaping the relationship between plant growth and the associated fungal diseases and pest populations. Recent predictions suggest that on a global scale, pests and diseases are moving to the poles at the rate of 3–5 km/year and the diversity of pest populations will also significantly change and have profound economic impact on staple food production systems [9, 10]. While these studies did not focus on mycotoxigenic fungal pathogens, this suggests a significant potential impact on mycotoxin contamination of staple foods/crops. Increases in pest reproduction rates would increase damage to ripening crops (during anthesis in wheat, and silking in maize) and facilitate more infection by mycotoxigenic fungi and contamination with mycotoxins.

The key environmental factors identified in these future changes therefore include: fluctuating humidity, including changes in rainfall patterns and extended periods of drought stress, increases and extreme temperature fluxes (global warming +3–5 °C), and an increase in CO_2 (two or three times present levels) [8, 11, 12].

This strong evidence showing a marked change in the environment in which staple crops will be grown in the next 10–25 years has increased the concern regarding the impact that predicted climate change scenarios may have on agriculture, mycotoxigenic fungal infection, and contamination with mycotoxins [11, 13–17].

Unfortunately, the current predictions and hypotheses about the real effect of climate change on mycotoxigenic fungi are based on historical or current climatic condition datasets which mostly consider interactions between water availability and temperature. Presently there are only very few studies examining the effect of 3-way interactions between the environmental factors identified (temperature, water availability, and CO_2) and what changes in terms of the ecophysiology of mycotoxigenic fungi and mycotoxin accumulation might have [16–18].

This chapter will consider the effects of the forecasted climate change factors of (1) a_w × temperature interactions and (2) three-way a_w × temperature × CO_2 interactions, on the ecophysiology of the main mycotoxigenic species. Potential future research areas which will need to be addressed will also be discussed.

4.2 Ecophysiological modifications on mycotoxigenic fungi under climate change conditions

4.2.1 Two-way a_w × temperature interactions

As discussed above, two of the key environmental factors identified in these future changes are fluctuating humidity and temperature. With regard to humidity, these fluctuations will vary from extreme rainfall to long periods of drought that will reduce the available water for living organisms. In soil, survival of water stress is predominantly determined by the total water potential and matric and solute components, whilst in food matrices the solute component and hence water activity (a_w) are more important [2]. On the other hand, as part of the forecasted environmental change, increases and extreme temperature fluxes have been predicted.

It is now accepted that these factors and their interactions are able to modulate germination, growth, sporulation, expression of genes involved in mycotoxins, and phenotypic mycotoxin production [19, 20] by mycotoxigenic fungi. Extreme drought episodes, desertification and fluctuations in wet/dry cycles will have an impact on their life cycles [21].

Magan et al. [11] reviewed the available ecological data on optimum and marginal interacting conditions of a_w × temperature for growth and mycotoxin production by several mycotoxigenic species. Effects on the growth of *Alternaria alternata, Fusarium proliferatum, F. verticillioides, F. culmorum, F. graminearum, Aspergillus westerdijkiae, A. carbonarius, A. flavus, Penicillium verrucosum, P. expansum* and production of alternariol, alternariol monomethyl ether, altenuene, fumonisins, trichothecenes, OTA, aflatoxin B_1 and patulin were considered when temperature was increased by either +3 or +5 °C at different water availabilities. These datasets have been combined with the available information from other publications [19, 22–26] and are shown in Tab. 4.2.

Only a_w × temperature interactions are considered. It is shown, however, that fungi would normally grow slower and in most cases produce similar or smaller amounts of mycotoxins under temperature and water stress. For some species however, such as *Aspergillus flavus*, which is able to grow and produce aflatoxin B_1 under high temperatures and to colonize maize efficiently under drought conditions, the forecasted conditions could become a problem particularly in the Mediterranean and other temperate regions [27].

Some examples of how extremely dry and warm conditions have affected the prevalence of mycotoxigenic fungi and mycotoxin contamination of staple food have been described. In 2003 and 2004 in northern Italy, and subsequently in the 2012 summer season, drought and extreme elevated temperatures (> 35 °C) resulted in a switch from *Fusarium verticillioides* and contamination with fumonisins to significant contamination of maize grain with *A. flavus* and aflatoxins, with later entry of aflatoxin M_1 into milk via the animal feed chain [27]. *A. flavus* has a wide range of temperature tolerance (19–35 °C) with about 28 °C being ideal for growth and 28–30 °C for aflatoxin

Tab. 4.2: Changes in growth and toxin production by *Alternaria* spp., *Fusarium* spp., *Aspergillus* spp., and *Penicillium* spp. as a result of increase in temperature of 3 or 5 °C under different water stress conditions.

	Growth				Toxins				
	a_w	μ_{max} range / Temp	μ_{+3}	μ_{+5}		a_w	τ_{max} range / Temp	τ_{+3}	τ_{+5}
Alternaria alternata	0.95	2–1 / 25	2–1	1–0.5	Altenuene	0.95	100–40 / 25	40–20	20–5
	0.90	0.1–0.5 / 25	0.5–0.1	NG		0.90	20–5 / 25	5–NP	NP
					Alternariol	0.95	500–100 / 25	40–20	20–5
						0.90	20–5 / 25	NP	NP
					Alternariol	0.95	400–100 / 25	100–10	NP
					monomethyl ether	0.90	100–10 / 25	NP	NP
Alternaria tenuissima	0.98	13.5–14 / 30	8–9.5	7.5–9	Altertoxin II	0.98	50–150 / 30	100–250	125–250
	0.95	5–6 / 30	4.5–5.5	4.5–5.5		0.95	600–2100 / 30	500–1000	400–1300
Fusarium proliferatum	0.95	4–3 / 28	3–2	2–1	Fumonisin	0.95	> 1000 / 20	1000–100	100–50
	0.90	0.5–0.1 / 28	NG	NG		0.93^a	50–10 / 15	50–10	50–10
Fusarium verticillioides	0.95	4–3 / 25	4–3	4–3	Fumonisin	0.95	10 000–1000 / 20	10 000–1000	1000–100
	0.90	0.5–0.1 / 25	0.5–0.1	NG		0.93^a	10 / 15	50–10	50–10
Fusarium culmorum	0.95	3–1 / 20	3–1	3–1	DON	0.95	1–0.25 / 20	0.25–0.1	0.1–0.01
	0.90	1–0.1 / 20	1–0.1	1–0.1		0.93^a	0.25–0.01 / 20	NP	NP
Fusarium graminearum	0.95	> 4 / 20	4–2	2–1	DON	0.95	1–0.1 / 20	1–0.1	0.1–0.01
	0.90	1–0.1 / 20	1–0.1	1–0.1		0.93^a	NP /	NP	NP
Fusarium langsethiae	0.98	4.6–5.1 / 25	4.6–4.1	3.6–3.1	T-2 + HT-2	0.98	10–11 / 25	13–14	14–15
	0.95	2.6–2.1 / 25	1.1–1.6	1.6–1.1		0.95	0–1 / 25	0–1	0–1

Tab. 4.2 (continued)

	Growth				Toxins				
	a_w	μ_{max} range / Temp	μ_{+3}	μ_{+5}		a_w	τ_{max} range / Temp	τ_{+3}	τ_{+5}
Aspergillus westerdijkiae	0.95	5–4.5 / 30	4.45–3.95	4.25–3.75	Ochratoxin A	0.95	1065.7–1014.9 / 30	719.4–685.1	488.5–465.3
	0.90	1.85–1.35 / 30	1.85–1.35	1.80–1.30		0.90	54.1–51.6 / 30	51.6–49.2	49.9–47.6
Aspergillus carbonarius	0.95	> 6 / 30	> 6	6–5	Ochratoxin A	0.95	2000–1500 / 20	1000–500	500–NP
	0.90	4–3 / 30	4–3	4–3		0.90	1000–500 / 20	500–NP	500–NP
Aspergillus flavus	0.95	6.9 / 35	5.6	5.0	Aflatoxin B_1	0.95	3082–2278 / 37	102–138	6.1–NP
	0.90	2.9 / 37	1.4	0.7		0.90	448.5–331.5 / 37	1–NP	NP
Penicillium verrucosum	0.97	2.2–2.4 / 25	1.8–1.9	0.9–1	Ochratoxin A	0.97	60–50/20 / 75	70– > 80	
	0.95	3–4 / 25	4–3	3–2		0.95	> 50 / 20	> 50	50–25
	0.90	2.5–1 / 25	2–1	2–0.5		0.90	50–30 / 20	50–30	50–30 (5–3[a])
Penicillium nordicum	0.97	1.4–1.5 / 20	1.5	2–1	Ochratoxin A	0.97	7–6 / 20	6–5	4–3
	0.94	1.8–2.0 / 20	1.8–1.7	1.4–1.7		0.94	7–6 / 20	7–6	7–6
	0.90	1.2–1.3 / 20	1.2–1.3	1.2–1.4		0.90	8–7 / 20	9–8	9–8
Penicillium expansum	0.98	9.1 / 20			Patulin	0.98	3.5–0.55 / 20	NS	NS
	0.95	6.9–5.5[a] / 25	6.9–5.5	6.9–5.5		0.95	NS /	NS	NS
	0.90	2.9–1.6[a] / 25	2.9–1.6	2.9–1.6		0.90	NS /	NS	NS

μ_{max}: Maximum growth rate (mm day^{-1}); μ_{+3}: Growth rate increasing 3 °C; μ_{+5} Growth rate increasing 5 °C;
τ_{max}: Maximum toxin production (µg ml^{-1}); τ_{+3}: Predicted toxin increase 3 °C; τ_{+5}: Predicted toxin increase 5 °C;
NG: no growth; NP: no toxin production; NS: not studied.

For *Aspergillus* and *Penicillium* species:
τ_{max}: Maximum toxin production (ng g^{-1}); τ+3, predicted toxin production with 3 °C increase; τ+5, predicted toxin production with 5 °C increase.

a Minimum water availability for toxin production.

production [19]. It has been shown that sporulation from sclerotia on maize stalks is maximized at 20–25 °C in a range of wet to dry conditions (0.995–0.90 a_w). Reduced sporulation occurred under drier conditions of ≤ 0.90 a_w [28]. Some strains of *A. flavus* can even grow at 0.73 a_w (in host tissue) and produce aflatoxins at 0.85 a_w, while in contrast, *F. verticillioides* growth is very marginal at 0.90 a_w and fumonisins are produced only at > 0.93 a_w [19]. *A. flavus*, a more xerotolerant mycotoxigenic species than *F. verticillioides*, was able to actively colonize ripening maize by outcompeting the more common non-xerophilic *Fusarium* species. Isolated strains were able to colonize ripening maize rapidly and produce both aflatoxins and cyclopiazonic acid [27]. Thus, because of the very dry and warm conditions in those years, *A. flavus* became a significant problem resulting in significant economic losses in the valuable and important cheese industry of the region. More recently, a Serbian maize survey in 2009–2011 identified no aflatoxin contamination. However, prolonged hot and dry weather in 2012 resulted in 69 % of samples being contaminated with aflatoxins [29]. Similarly, in Hungary, it has also been shown that an increase in aflatoxins may be due to climate change conditions [30]. However, there are only a few more concrete examples of such incidences where climate change factors have been implicated.

Magan et al. [11] suggested that drought conditions favored by climate change may result in xerophilic fungi such as *Wallemia sebi*, *Xeromyces bisporus*, and *Chrysosporium* species becoming more important as colonizers of food commodities. These species are able to grow under very dry conditions (0.65–0.75 a_w) where there is much less competition from the majority of mesophilic fungi [2, 31]. This potential change in the food ecosystem may pose a risk, since *W. sebi* can produce metabolites such as walleminol and walleminone which can be toxic to animals and humans [32]. Recent research by Leong et al. [33] suggests that there are competitive interactions between these xerophilic fungi in extreme dry conditions and that secondary metabolites may play a role.

Another group of species which could benefit from these warmer and drier conditions could be species from the *Aspergillus* section *Nigri*. Their optimal growth conditions have been shown to be between 0.95–0.99 a_w and 30–37 °C and optimal OTA production between 0.95–0.99 a_w and 15–20 °C [33, 35–38]. Among these species *A. carbonarius* has been identified as being able to produce high OTA amounts in a wide range of fruits and dried fruits. However, other species from the *A. niger* aggregate have also been reported to be OTA producers. Among these, for example, *A. tubingensis*, was described as an OTA producer by Medina et al. [39], although it has a different growth and toxin production pattern. In strains isolated from Morocco, the optimal temperature and a_w conditions for OTA production was in the range of 25–30 °C and 0.95–0.99 a_w for *A. carbonarius* and 30–37 °C and 0.90–0.95 a_w for *A. niger* and *A. tubingensis* [40]. Interestingly, Chiotta et al. [41] studied the effect of interactions between *A. tubingensis* and *A. carbonarius* strains and found that OTA production by the latter species was mainly dependent on temperature. At 35 °C, *A. tubingensis* reduced OTA production by *A. carbonarius*, an effect not observed at 20 °C. Recently, García-Cela et al. [42] concluded that climate change prediction points to drier and hotter climatic scenarios where *A. tubingensis* and *A. niger* could be even more prevalent than *A. carbonarius*, since they are better adapted to extreme high temperatures and drier conditions. Gil-Serna et al. [43] also studied growth and OTA production by *A. steynii* and *A. westerdijkiae* in a green-coffee based medium under different environmental conditions. They demonstrated that the optimal conditions of both species to grow and produce OTA were at 28–32 °C and a_w levels > 0.95 a_w. *A. steynii* was able to grow and produce OTA over a wider set of conditions than *A. westerdijkiae*. Thus, climate change might result in a shift of the main source of OTA in several products since drier and warmer conditions will benefit the colonization and OTA production by species from the *Aspergillus* section *Nigri* species.

There might also be changes in mycotoxin prevalence among the same species. As an example, *Alternaria alternata* is able to produce a range of secondary metabolites. Among others, it produces mycotoxins such as alternariol (AOH) and alternariol monomethyl ether (AME). While maximum AOH production is reached at 21 °C and 0.95 a_w, AME production was maximized at the same a_w levels but at much warmer conditions of 35 °C. Thus, increases in temperature may equate to shifts from AOH to AME under the forecasted conditions [44]. Schmidt-Heydt et al. [45] also demonstrated that temperature affects mycotoxin production pattern of *A. parasiticus*; with an increasing temperature (> 30 °C), there is a shift from aflatoxin G_1 to aflatoxin B_1. However, these studies exclude interaction with CO_2 which is necessary to examine the impact of predicted climate change scenarios in more detail.

4.2.2 Three-way a_w × temperature × CO_2 interactions

The addition of CO_2 as a potential factor influencing mycotoxigenic fungi seems to be more than plausible. Doubling its current concentration around 2050 and tripling it around 2100, CO_2 will have a significant effect on all living species around the globe and plants and fungal species will be no exception. Considering the effect on plants, photosynthesis, leaf area, plant height, total biomass and crop yield, sugar and starch content, water-use efficiency, growth, and yield are generally increased in the presence of higher levels of CO_2 [46].

In contrast, fungi, including mycotoxigenic fungi, are able to withstand quite high concentrations of CO_2 [47]. Thus, tolerance of climate change factors where perhaps a tripling of the existing CO_2 is predicted to occur may not be problematic for the growth of mycotoxigenic fungi.

The available evidence also shows that the predicted climate change conditions are expected to alter the potential risk of plant diseases [48, 49] and elicit changes in host and pathogen interactions, such as an increase or decrease in disease symptoms, infection or pest fecundity [46] which will affect crop yield and quality [50, 51]. Carbon dioxide may also influence plant resistance [52].

Although there are currently serious concerns regarding food security and food safety accompanying crop productivity and mycotoxin contamination in the context of global climate change, our understanding of how elevated temperature × water stress × elevated CO_2 interactions are able to modulate crop plants, mycotoxigenic fungi, and the plant/fungus interaction is still very limited.

In the last two decades, research on the impact of CO_2 focused predominantly on its use as part of a modified atmosphere storage system to control growth and damage to commodities during storage. With regard to mycotoxigenic fungi, it has been shown that they are very tolerant of between 25–75 % CO_2, regardless of a_w or temperature [47]. The approach has been to reduce the O_2 concentration to < 2 % and increase CO_2 to > 50–60 %. However, many fungi are microaerophilic and will still be able to grow slowly under such conditions in storage systems and packaging [53, 54]. Wilson and Jay [55] tested a high CO_2 treatment (61.7 % CO_2 balanced with O_2 and N_2) on moist maize and found that *A. flavus* growth was visible after 4 weeks at 27 °C. Contamination with aflatoxins at elevated CO_2 was lower than that in air. Although no mycotoxin analyses were carried out, Magan and Lacey [56] examined the interactions between reduced O_2 concentration (21–0.14 %) or elevated CO_2 (5–15 %) balanced with nitrogen and interactions with temperature (14, 23 °C) and water stress (0.98–0.85 a_w) for a range of 6 phyllosphere cereal fungi and 10 spoilage *Aspergillus* and *Penicillium* species on wheat-based media. For example, the latent periods prior to growth (lag times) for *Alternaria alternata* and *Fusarium culmorum* were significantly increased by > 5 % CO_2, regardless of a_w or temperature. Growth of *A. alternata* at 23 °C was inhibited by > 5 % CO_2 at 0.98 and 0.95 a_w. On the other hand, at 0.90 a_w growth was stimulated by 5 % and 10 % CO_2. This also occurred at 14 °C and 0.95 a_w. In contrast, growth of *F. culmorum* was inhibited significantly by > 5 % CO_2 at 0.98 and 0.95 a_w at both temperatures, whilst appearing to be unaffected at 0.90 a_w [56, 57].

The effects of elevated CO_2 on *A. ochraceus* (= *A. westerdijkiae*) and *P. verrucosum* growth and OTA production have also been determined [58, 59]. Studies by Pateraki et al. [53] suggested that up to 50 % CO_2 had only a slight impact of OTA production by *A. carbonarius* over a range of a_w conditions, with this being a more important factor than CO_2. Samapundo et al. [60] found that fumonisin B_1 production by *Fusarium* section *Liseola* species (*F. verticillioides*, *F. proliferatum*) was inhibited by 30 % CO_2 at 0.984 a_w.

In both *in vitro* and *in situ* studies, Giorni et al. [54] demonstrated that CO_2 × a_w interactions significantly decreased the ability of *A. flavus* to grow and colonize maize grain. The use of modified atmospheres at 25 % and 50 % CO_2 resulted in about 30–35 % inhibition of growth (CFU g^{-1} grain). Exposure to 75 % CO_2 resulted in > 50 % inhibition of growth and aflatoxin B_1 production, regardless of the a_w level. Other studies of modified atmospheres with different CO_2 levels balanced with O_2 and N_2 showed that *A. flavus* grew on wheat and rye bread with up to 75 % CO_2 [61]. Exposure to 70 % CO_2 at 0.80 a_w prevented spoilage of bakery products; however, when a_w was 0.85 or 0.90 spoilage was delayed [62]. These previous studies demonstrate that very high CO_2 concentrations are needed to control growth and associated mycotoxin production by the main mycotoxigenic fungi.

Medina et al. [16] recently examined the effect of current environmental conditions and compared them with forecasted conditions. Temperatures were changed from 34 to 37 °C, drought stress was imposed, and CO_2 was increased from 350 to 650 and 1000 ppm. The effects on growth of *A. flavus* and aflatoxin B_1 production were examined. This was the first time a combination of expected climate change factors were used to establish the potential effects of these scenarios on the ecophysiology of mycotoxigenic fungi. The results showed that for growth of *A. flavus* there was relatively little effect of these interacting conditions (Fig. 4.2 (a)). In contrast, the three-way interacting conditions had a more profound effect on aflatoxin B_1 production. This clearly demonstrates a stimulation of aflatoxin B_1 production under slightly elevated CO_2 conditions, especially under drought stress at 37 °C and 650 and 1000 ppm CO_2 exposure (Fig. 4.2 (b)). It seems that the interactions between all three factors are crit-

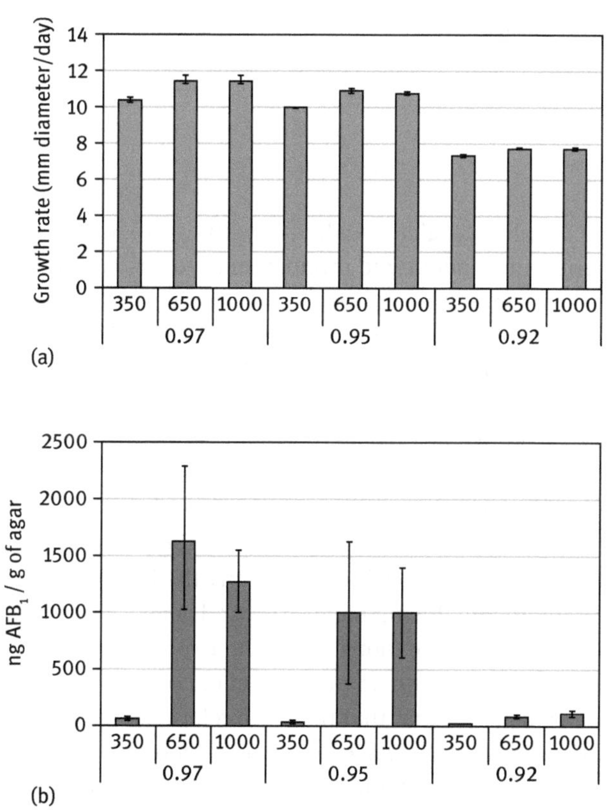

Fig. 4.2: Effect of water activity (0.97, 0.95, 0.92) × elevated CO_2 levels (350, 650, 1000 ppm) on (a) relative growth rates; and (b) aflatoxin B_1 production, of an *A. flavus* strain on a conducive YES medium at 37 °C. Bars indicate standard errors. Adapted from Medina et al. [16].

ical in the impact that slightly elevated CO_2 has. This is clear from the results obtained at 0.92 and 0.95 a_w × 37 °C and 650 or 1000 ppm CO_2, where a statistically significant increase in aflatoxin B_1 was observed.

Vaughan et al. [18] recently investigated the impact of elevated CO_2 on the interaction between maize and *F. verticillioides*. They found that elevated CO_2 of approx. 800 µmol CO_2 mol^{-1} air (≈ double current CO_2) increased maize susceptibility to *F. verticillioides* colonization. Interestingly, fumonisin production was not affected by these interactions. They also found that fumonisin production was not proportional to the increase in the biomass of *F. verticillioides*, and the amount of fumonisin produced per unit pathogen was reduced in elevated CO_2. Their study also showed that physiological impacts on maize agronomy were evident when grown under the climate change treatments used, including 28 °C day/25 °C night temperature cycles, 500 µmol m^{-2} s^{-1} photosynthetic photo flux density 12 h photoperiod and 400 µmol CO_2 mol^{-1} air (≈ current CO_2) and the other at 800 µmol CO_2 mol^{-1} air (≈ double current CO_2). These physiological effects on the maize plant could further impact on infection and contamination with mycotoxins, especially during silking.

4.3 Climate change impact on mycotoxin gene cluster expression and its relationship to growth and toxin production.

Integration of the correlation of ecophysiological conditions with expression of specific mycotoxin biosynthesis genes and phenotypic production of toxins could help improve knowledge and understanding of how climate change conditions influence the growth and regulation of mycotoxins. The clusters of genes involved in the production of key toxic secondary metabolites including aflatoxins, OTA, trichothecenes such as deoxynivalenol, and fumonisins have been largely identified [63, 64]. The genes involved in mycotoxin production are often clustered together and key marker genes in the pathways have been identified. Most studies have focused on research into how key abiotic factors, such as a_w and temperature, affect expression of the genes involved in mycotoxin biosynthesis. Practically none have examined the effects of CO_2.

The aflatoxin cluster of genes consists of 25 genes, including the key regulatory genes (*aflR* and *aflS*) and a series of up and downstream structural genes. It has been shown that both a_w and temperature modifications affect the expression of these clusters of genes, relative growth rate, and aflatoxin production in both *A. flavus* and *A. parasiticus* [45, 65]. It has been suggested that that there is a relationship between aflatoxin biosynthetic genes (*aflS/aflR* ratio and *aflD* genes) and aflatoxin B_1 production when exposed to elevated temperature and drought stress conditions [66]. However, the effect of CO_2 was not included in these studies. More recent detailed studies using a mycotoxin microarray [67] have been useful in elucidating the relationship between both key regulatory genes and the structural genes and interacting conditions of a_w × temperature [66]. As the a_w increased, the expression of *aflD* was reduced; the regulatory genes *aflR* and *aflS* were less sensitive to a_w having a similar sensitivity. As the temperature was increased, the expression of *aflD* and *aflR* was reduced and that of *aflS* increased. A slight change in temperature caused a large change in expression of *aflD* and *aflR*. Recently, Medina et al. [16] studied the effect of temperature × a_w × CO_2 on the relative expression of the mycotoxin biosynthetic genes, specifically, on structural *aflD* and the regulatory *aflR* genes involved in aflatoxin B_1 production. There was a stimulation of *aflD* gene expression when temperature increased from 34 to 37 °C and at 650 and 1000 ppm CO_2 at both 0.97 and 0.95 a_w. For the regulatory gene *aflR* expression, there was an increase at either 650 or 1000 ppm CO_2 exposure, especially at 0.95 and 0.92 a_w and at 37 °C. In addition, it was demonstrated that these three-way interactions of climate change factors stimulated the relative expression of genes involved in the biosynthesis of aflatoxin production, resulting in an increase in phenotypic aflatoxin B_1 being produced.

The fumonisin cluster of genes for *F. verticillioides* has 17 genes, among them the *FUM1* and *FUM21* genes are considered important, while *F. sporotrichioides* and *F. oxysporum* have 18 [63]. Medina et al. [68] integrated toxin gene expression, growth and fumonisin B_1 and B_2 by a strain of *F. verticillioides* under different a_w × temperature combinations. This study showed that activation of the expression of nine biosynthetic cluster genes (*FUM1, FUM7, FUM10, FUM11, FUM12, FUM13, FUM14, FUM16,* and *FUM19*) was related to growth and phenotypic FB_1 and FB_2 production. Thus, for example, *FUM11* and *FUM13* gene expression was stimulated when the temperature was increased from 20 to 25 °C and at 0.93 a_w (Fig. 4.3). For *FUM1* gene expression,

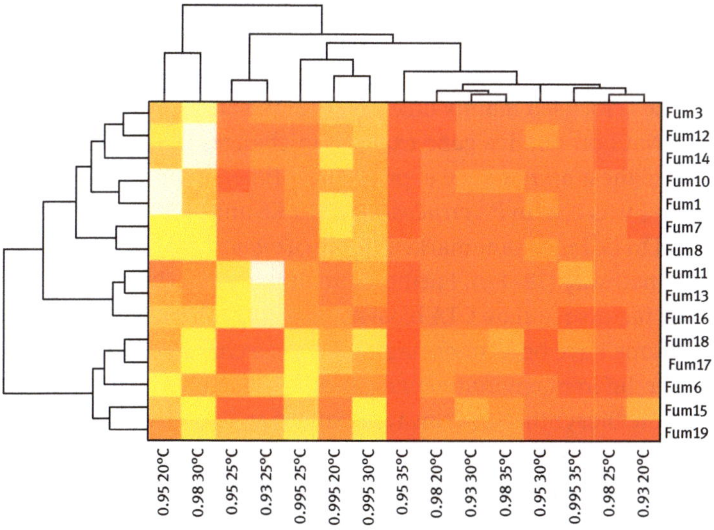

Fig. 4.3: Heat map: relative expression of 15 genes from the fumonisins gene cluster in relation to a_w × temperature conditions. Data obtained using a mycotoxin microarray [67]. Adapted from Medina et al. [68].

there was an increase at 0.95 a_w and at 20 °C. Lazzaro et al. [69] also studied the effect of temperature and a_w on two fumonisin biosynthetic genes (*FUM1* and *FUM21*) and fumonisin production by *F. verticilloides*. It was shown that temperature only significantly affected *FUM21* expression, with the optimum at 25 °C, while a_w did not have a significant effect.

Although some of the genes involved in the trichothecene pathway such as *Tri7* and *Tri13* [70], and *Tri12* [71] have been used as targets for developing PCR assays for characterization of chemotypes of different *Fusarium* species, the *Tri5* gene has been used to study its expression in relation to deoxynivalenol production by mycotoxigenic *Fusarium* species under different temperatures and a_w conditions. Marin et al. [72] investigated the effect of a_w and temperature on *Tri5* gene expression by *F. graminearum*. They found that the expression of the *Tri5* gene was constant in all conditions tested, although some induction was observed between 20 and 30 °C. Schmidt-Heydt et al. [73] used a microarray to check the expression of six genes (*Tri4, Tri5, Tri6, Tri10, Tri12*, and *Tri13*) involved in trichothecene biosynthesis to model the relationship between environmental factors, transcriptional genes, and deoxynivalenol mycotoxin production by strains of *F. graminearum* and *F. culmorum*. This study showed that there were different patterns of gene expression depending on abiotic conditions and species. The expression data for the strain of *F. culmorum* were relatively much higher than for the *F. graminearum* strain. It was observed that at warmer conditions (25 °C) and high a_w levels, but also at low a_w, the biosynthetic genes were more expressed. Therefore, these species could be considered well-adapted in predicted climate change scenarios resulting in a potential high risk source of trichothecenes contamination.

Regarding the OTA biosynthesis pathway, OTA producing *Penicillia* have been found to have at least 4 genes involved in OTA production with the *otapks* and *otanps* genes key biosynthetic pathway genes [74]. Although both key genes have been used for expression studies, these focused mainly on evaluating the effect of preservatives or ionic solutes such as NaCl on relative gene expression changes [25, 75, 76]. Some studies showed that the highest *otapks* gene expression in *P. nordicum* was at 15 and 25 °C [77], and when the a_w was 0.98 in *P. verrucosum*. The PKS and P450 related genes have recently been found to be important markers for *A. carbonarius* and *A. westerdijkiae* in the biosynthetic pathway [78–80]. Furthermore, it has been shown that VeA and LaeA transcriptional factors regulate OTA biosynthesis in *A. carbonarius* [81]. Unfortunately no investigations have been performed to analyze changes in expression of the former genes under different environmental conditions.

The genetic basis of the biosynthetic pathway of patulin in *P. expansum* has been recently characterised [82]. It consists of 15 genes with the *patK* and *patN* genes being necessary for patulin production [83]. So far no studies have been carried out to evaluate the influence of environmental factors and/or climate change scenarios on the expression of genes involved in patulin production.

Although a number of relevant investigations have examined the effect of environmental factors (a_w, temperature) on biosynthetic genes, it is of paramount importance to continue research on the influence of biotic and abiotic factors on different genes which have not been studied previously and which may provide better insights into the impact of slightly elevated CO_2 on the physiology of the plant and the interface with infection/colonization by different mycotoxigenic fungi. Quantification of key biosynthetic toxin genes under different climate change scenarios in the host/pathogen interface could be a valuable tool for gaining knowledge of the ecophysiological basis for mycotoxin biosynthesis in relation to climate change factors and also facilitate better control strategies to avoid mycotoxin contamination of staple food commodities.

4.4 Conclusions

Because of their ability to adapt to change, fungal species, especially mycotoxigenic ones, may become a primary concern in the coming 20–25 years. They exert an as yet incompletely understood mechanism of metabolism with a high degree of plasticity which allows them to shape the metabolic responses to interacting environmental factors. Thus, precise forecasting with regard to mycotoxigenic fungal populations and mycotoxin contamination and prevalence in the coming years remains an area for research. This chapter points out that interactions between CO_2, temperature, and a_w may have differential effects which are related to both plant physiology and fungal pathogenic species involved. While for some fungal species growth or mycotoxin production remains similar under the forecasted conditions, for others environmental changes may have significant effects, e.g. increasing toxin production or leading to a switch in the major mycotoxin produced or the ratio of different mycotoxins. Perhaps the addition of the new findings with regard to toxin production under elevated CO_2 and temperature conditions discussed in this chapter would reshape the outcome of a recent report published by EFSA with regard to relative risk of contamination with aflatoxin B_1 in maize, rice, and wheat [8]. Nonetheless, much more data is required to enable a better understanding of fungal and plant ecophysiology and the pathogen/host interface and to improve the potential for making more accurate and relevant global predictions.

Furthermore, toxin production/mycotoxin biosynthetic gene expression are not related to growth per se, so more research is needed to establish the potential effect of these factors and to understand how gene expression is related to phenotypic toxin production. Considering all the information, several questions remain unanswered and research efforts are needed to improve the current understanding. Are the new climate change factors going to change toxin production patterns? Will other mycotoxins now considered secondary become more abundant and thus more important in the future? Are the current control/mitigation strategies going to be effective in the future? Will the agricultural practices used currently in order to minimize toxin contamination be suitable when marked environmental changes become standard? Are mycotoxigenic fungal populations going to shift their location in the coming years?

More research is urgently required to address these key questions and allow effective prediction of the relative level of risk of different mycotoxins in economically important staple food crops.

Dedication

The authors wish to dedicate this chapter to the memory of Dr. Sejakhosi Alexis Mohale (University of Lesotho), who completed his PhD at Cranfield in 2013 on "Biological control of aflatoxins using atoxigenic strains of *A. flavus*", and who sadly passed away in August 2014.

References

[1] Rousseau JJ. Meditations (Reveries) of the solitary walker. London, UK: Penguin Books; 1995.

[2] Magan N. Fungi in extreme environments. In: Kubicek CP, Druzhinina IS, eds. *Environmental and Microbial Relationships*, 2nd ed. Berlin, Germany: Springer Verlag; 2007. p. 85–103.

[3] Williams JP, Hallsworth JE. Limits of life in hostile environments: no barriers to biosphere function? *Environ Microbiol* 2009;11:3292–3308.

[4] European Commission. *Commission regulation (EC) 1881/2006 of 19 December 2006 setting maximum levels for certain contaminants in foodstuffs*. OJEU 2006;L364:5–24.

[5] European Commission. *Adapting to climate change in Europe–options for EU action. Green paper from the Commission to the Council, the European Parliament, the European Economic and Social Committee of the Regions*. COM (2007) 354, Brussels, Belgium: European Commission; 2007. (Accessed May 5, 2014, at http://eur-lex.europa.eu/legal-content/EN/TXT/?uri=CELEX:52007DC0354)

[6] IPCC. *Climate change 2007: Synthesis report. Intergovernmental panel on climate change*. Geneva, Switzerland; 2007. (Accessed May 5, 2014, at http://www.ipcc.ch/publications_and_data/publications_ipcc_fourth_assessment_report_synthesis_report.htm)

[7] Solomon S, Alley RB, Bertsen T et al. Technical summary. In: Solomon S, Qin D, Manning M et al, eds. *Climate Change 2007: The physical science basis – Contribution of Working Group1 to the Fourth Assessment Report of the Intergovernmental Panel on Climate Change*. Cambridge, UK: Cambridge University Press; 2007. p. 19–91.

[8] Battilani P, Rossi V, Giorni P et al. *Modelling, predicting and mapping the emergence of aflatoxins in cereals in the EU due to climate change*. European Food Safety Authority (EFSA), Supporting publication, 2012. (Accessed May 5, 2014, at http://www.efsa.europa.eu/en/supporting/pub/223e.htm)

[9] Bebber DP, Ramotowski MAT, Gurr SJ. Crop pests and pathogens move poleward in a warming world. *Nature Clim Change* 2013;3:985–988.

[10] Bebber DP, Holmes T, Gurr SJ. The global spread of crop pests and pathogens. *Global Ecol Biogeogr* 2014;23:1398–1407.

[11] Magan N, Medina A, Aldred D. 2011. Possible climate-change effects on mycotoxin contamination of food crops pre- and postharvest. *Plant Pathol* 2011;60:150–163.

[12] Beddington J, Asaduzzaman M, Clark M et al. *Achieving food security in the face of climate change: Final report from the Commission on Sustainable Agriculture and Climate Change.*

CGIAR Research Program on Climate Change, Agriculture and Food Security (CCAFS). Copenhagen, Denmark; 2012. (Accessed May 5, 2014, at http://ccafs.cgiar.org/commission/reports#.VOTbnCSDOUk)

[13] Paterson R, Lima N. How will climate change affect mycotoxins in food? *Food Res Int* 2010;43:1902–1914.

[14] Paterson R, Lima N. Further mycotoxin effects from climate change. *Food Res Int* 2011;44: 2555–2566.

[15] Wu F, Bhatnagar D, Bui-Klimke T et al. Climate change impacts on mycotoxin risks in US maize. *World Mycotoxin J* 2011;4:79–93.

[16] Medina A, Rodriguez A, Magan N. Effect of climate change on *Aspergillus flavus* and aflatoxin B_1 production. *Front Microbiol* 2014;5:348. doi: 10.3389/fmicb.2014.00348.

[17] Medina A, Rodríguez A, Sultan Y, Magan N. Climate change factors and *Aspergillus flavus*: effects on gene expression, growth and aflatoxin production. *World Mycotoxin J* 2015;8:171–179.

[18] Vaughan MM, Huffaker A, Schmelz EA et al. Effects of elevated [CO_2] on maize defence against mycotoxigenic *Fusarium verticillioides*. *Plant Cell Environ* 2014;37:2691–2706.

[19] Sanchis V, Magan N. Environmental conditions affecting mycotoxins. In: Magan N, Olsen M, eds *Mycotoxins in food: detection and control*. Abington, Cambridge, UK: Woodhead Publishing Ltd and CRC Press LLC; 2004. p. 174–189.

[20] Magan N, Geisen R, Schmidt-Heydt M, Medina A, Parra R, Abdel-Hadi A. A systems approach to integrating molecular, ecophysiological data and phenotypic data for a better understanding of mycotoxin contamination. *J Plant Pathol* 2012;94:39.

[21] Garcia D, Barros G, Chulze S, Ramos AJ, Sanchis V, Marín S. Impact of cycling temperatures on *Fusarium verticillioides* and *Fusarium graminearum* growth and mycotoxins production in soybean. *J Sci Food Agric* 2012;92:2952–2959.

[22] Medina A, Magan N. Comparisons of water activity and temperature impacts on growth of *Fusarium langsethiae* strains from northern Europe on oat-based media. *Int J Food Microbiol* 2010;142:365–369.

[23] Medina A, Magan N. Temperature and water activity effects on production of T-2 and HT-2 by *Fusarium langsethiae* strains from north European countries. *Food Microbiol* 2011;28: 392–398.

[24] Patriarca A, Medina A, Fernández Pinto V, Magan N. Temperature and water stress impacts on growth and production of altertoxin-II by strains of *Alternaria tenuissima* from Argentinean wheat. *World Mycotoxin J* 2014;7:329–334.

[25] Rodríguez A, Medina A, Córdoba JJ, Magan N. The influence of salt (NaCl) on ochratoxin A biosynthetic genes, growth and ochratoxin A production by three strains of *Penicillium nordicum* on a dry-cured ham-based medium. *Int J Food Microbiol* 2014;178:113–119.

[26] Rodríguez A, Capela D, Medina A, Córdoba JJ, Magan N. Relationship between ecophysiological factors, growth and ochratoxin A contamination of dry-cured sausage based matrices. *Int J Food Microbiol* 2015;194:71–77.

[27] Giorni P, Magan N, Pietri A, Bertuzzi T, Battilani P. Studies on *Aspergillus* section *Flavi* isolated from maize in northern Italy. *Int J Food Microbiol* 2007;113:330–338.

[28] Giorni P, Camardo Leggieri M, Magan N, Battilani P. Comparison of temperature and moisture requirements for sporulation of *Aspergillus flavus* sclerotia on natural and artificial substrates. *Fungal Biol* 2012;116:637–642.

[29] Kos J, Mastilović J, Janić Hajnal E, Šarić B. Natural occurrence of aflatoxins in maize harvested in Serbia during 2009–2012. *Food Control* 2013;34:31–34.

[30] Dobolyi C, Sebők F, Varga J et al. Occurrence of aflatoxin producing *Aspergillus flavus* isolates in maize kernel in Hungary. *Acta Aliment Hung* 2013;42:451–459.

[31] Magan N, Aldred D. Environmental fluxes and fungal interactions: maintaining a competitive edge. In: van West P, Avery S, Stratford M, eds *Stress in Yeasts and Filamentous Fungi*. Amsterdam, the Netherlands: Elsevier Ltd; 2008. p. 19–35.
[32] Piecková E, Kunová Z. Indoor fungi and their ciliostatic metabolites. *Ann Agric Environ Med* 2002;9:59–63.
[33] Leong S, Pettersson OV, Rice T, Hocking AD, Schnurer J. The extreme xerophilic mould *Xeromyces bisporus* – growth and competition at various water activities. *Int J Food Microbiol* 2011;145:57–63.
[34] Bellí N, Ramos AJ, Sanchís V, Marín S. Incubation time and water activity effects on ochratoxin A production by *Aspergillus* section *Nigri* strains isolated from grapes. *Lett Appl Microbiol* 2004;38:72–77.
[35] Esteban A, Abarca ML, Bragulat MR, Cabañes FJ. Effects of temperature and incubation time on production of ochratoxin A by black aspergilli. *Res Microbiol* 2004;155:861–866.
[36] Mitchell D, Parra R, Aldred D, Magan N. Water and temperature relations of growth and ochratoxin A production by *Aspergillus carbonarius* strains from grapes in Europe and Israel. *J Appl Microbiol* 2004;97:439–445.
[37] Astoreca AL, Magnoli CE, Dalcero A. Ecophysiology of *Aspergillus* section *Nigri* species potential ochratoxin A producers. *Toxins* 2010;2:2593–2605.
[38] Astoreca AL, Barberis CL, Magnoli CE, Dalcero A. Growth and ochratoxin A production by *Aspergillus niger* group strains in coffee beans in relation to environmental factors. *World Mycotoxin J* 2010;3:59–65.
[39] Medina A, Mateo R, López-Ocaña L, Valle-Algarra FM, Jiménez M. Study of Spanish grape mycobiota and ochratoxin A production by Isolates of *Aspergillus tubingensis* and other members of *Aspergillus* section *Nigri*. *Appl Environ Microbiol* 2005;71:4696–4702.
[40] Selouane A, Bouya D, Lebrihi A, Decock C, Bouseta A. Impact of some environmental factors on growth and production of ochratoxin A of/by *Aspergillus tubingensis*, *A. niger*, and *A. carbonarius* isolated from moroccan grapes. *J Microbiol* 2009;47:411–419.
[41] Chiotta ML, Sosa DM, Ponsone ML, Chulze SN. Effect of water activity and temperature on growth of *Aspergillus carbonarius* and *Aspergillus tubingensis* and their interactions on ochratoxin A production. *World Mycotoxin J* 2015;8:99–105.
[42] García-Cela E, Crespo-Sempere A, Ramos AJ, Sanchis V, Marin S. Ecophysiological characterization of *Aspergillus carbonarius*, *Aspergillus tubingensis* and *Aspergillus niger* isolated from grapes in Spanish vineyards. *Int J Food Microbiol* 2014;173:89–98.
[43] Gil-Serna J, Vázquez C, García Sandino F, Márquez Valle A, González-Jaén MT, Patiño B. Evaluation of growth and ochratoxin A production by *Aspergillus steynii* and *Aspergillus westerdijkiae* in green-coffee based medium under different environmental conditions. *Food Res Int* 2014;61:127–131.
[44] Vaquera S, Patriarca A, Fernández Pinto V. Water activity and temperature effects on growth of *Alternaria arborescens* on tomato medium. *Int J Food Microbiol* 2014;185:136–139.
[45] Schmidt-Heydt M, Rüfer CE, Abdel-Hadi A, Magan N, Geisen R. The production of aflatoxin B_1 or G_1 by *Aspergillus parasiticus* at various combinations of temperature and water activity is related to the ratio of *aflS* to *aflR* expression. *Mycotoxin Res* 2010;26:241–246.
[46] Eastburn DM, DeGennaro MM, DeLucia EH, Dermody O, McElrone AJ. Elevated atmospheric carbon dioxide and ozone alter soybean diseases at SoyFACE. *Glob Change Biol* 2010;16: 320–330.
[47] Magan N, Aldred D. Post-harvest control strategies: minimizing mycotoxins in the food chain. *Int J Food Microbiol* 2007;119:131–139.
[48] Chakraborty S, Newton AC. Climate change, plant diseases, and food security: an overview. *Plant Pathol* 2011;60:2–14.

[49] Juroszek P, von Tiedemann A. Potential strategies and future requirements for plant disease management under a changing climate. *Plant Pathol* 2011;60:100–112.

[50] Ziska LH, Morris CF, Goins EW. Quantitative and qualitative evaluation of selected wheat varieties released since 1903 to increasing atmospheric carbon dioxide: can yield sensitivity to carbon dioxide be a factor in wheat performance? *Glob Change Biol* 2004;10:1810–1819.

[51] Dixon GR. Climate change impact on crop growth and food production, and plant pathogens. *Can J Plant Pathol* 2012;34:362–379.

[52] Plazek A, Hura K, Rapacz H, Zur I. The influence of ozone fumigation on metabolic efficiency and plant resistance to fungal pathogens. *J Appl Bot* 2001;75:8–13.

[53] Pateraki M, Dekanea A, Mitchell D, Lydakis D, Magan N. Influence of sulphur dioxide, controlled atmospheres and water availability on in vitro germination, growth and ochratoxin A production by strains of *Aspergillus carbonarius* from grapes. *Postharvest Biol Tec* 2007;44:141–149.

[54] Giorni P, Battilani P, Pietri A, Magan N. Effect of a_w and CO_2 level on *Aspergillus flavus* growth and aflatoxin production in high moisture maize post-harvest. *Int J Food Microbiol* 2008;122:109–113.

[55] Wilson DM, Jay E. Influence of modified atmosphere storage on aflatoxin production in high moisture corn. *Appl Microbiol* 1975;29:224–228.

[56] Magan N, Lacey J. Effect of gas composition and water activity on growth of field and storage fungi and their interactions. *T Brit Mycol Soc* 1984;82:305–314.

[57] Magan N. Studies on the mycoflora of wheat grain: Ecology of the fungi and effects of fungicides. Reading, UK: Reading University; PhD thesis 1982.

[58] Paster N, Lisker N, Chet I. Ochratoxin A production by *Aspergillus flavus* Wilhelm grown under controlled atmospheres. *Appl Environ Microbiol* 1983;45;1136–1139.

[59] Cairns-Fuller V, Aldred D, Magan N. Water, temperature and gas composition interactions affect growth and ochratoxin A production by isolates of *Penicillium verrucosum* on wheat grain. *J Appl Microbiol* 2005;99:1215–1221.

[60] Samapundo S, De Meulenaer B, Atukwase A, Debevere J, Devlieghere F. The influence of modified atmospheres and their interaction with water activity on the radial growth and fumonisin B_1 production of *Fusarium verticillioides* and *F. proliferatum* on corn. Part 1: The effect of initial headspace carbon dioxide concentration. *Int J Food Microbiol* 2007;114:160–167.

[61] Suhr KI, Nielsen PV. Inhibition of fungal growth on wheat and rye bread by modified atmosphere packaging and active packaging using volatile mustard essential oil. *J Food Sci* 2005;70:37–44.

[62] Guynot ME, Marín S, Sanchis V, Ramos AJ. Modified atmosphere packaging for prevention of mold spoilage of bakery products with different pH and water activity levels. *J Food Protect* 2003;66:1864–1872.

[63] Brown DW, Proctor RH. *Fusarium*: genomics, molecular and cellular biology. Portland, OR, USA: Caister Academic Press; 2013.

[64] Yu J, Chang PK, Ehrlich KC et al. Clustered pathway genes in aflatoxin biosynthesis. *Appl Environ Microbiol* 2004;70:1253–1262.

[65] Schmidt-Heydt M, Abdel-Hadi A, Magan N, Geisen R. Complex regulation of the aflatoxin biosynthesis gene cluster of *Aspergillus flavus* in relation to various combinations of water activity and temperature. *Int J Food Microbiol* 2009;135:231–237.

[66] Abdel-Hadi A, Schmidt-Heydt M, Parra R, Geisen R, Magan N. A systems approach to model the relationship between aflatoxin gene cluster expression, environmental factors, growth and toxin production by *Aspergillus flavus*. *J R Soc Interface* 2012;9:757–767.

[67] Schmidt-Heydt M, Geisen R. A microarray for monitoring the production of mycotoxins in food. *Int J Food Microbiol* 2007;117:131–140.

[68] Medina A, Schmidt-Heydt M, Cárdenas-Chávez DL, Parra R, Geisen R. Magan N. Integrating toxin gene expression, growth and fumonisin B_1 and B_2 production by a strain of *Fusarium verticillioides* under different environmental factors. *J R Soc Interface* 2013;10:20130320. doi: 10.1098/rsib.2013.0320.

[69] Lazzaro I, Susca A, Mulè G et al. 2012. Effects of temperature and water activity on *FUM2* and *FUM21* gene expression and fumonisin B production in *Fusarium verticillioides. Eur J Plant Pathol* 2012;134:685–695.

[70] Chandler EA, Simpson DR, Thomsett MA, Nicholson P. Development of PCR assays to *Tri7* and *Tri13* trichothecene biosynthetic genes, and characterisation of chemotypes of *Fusarium graminearum*, *Fusarium culmorum* and *Fusarium cerealis*. *Physiol Mol Plant Pathol* 2003;62:355–367.

[71] Nielsen LK, Jensen JD, Rodríguez A, Jørgensen LN, Justesen AF. *Tri12* based quantitative real-time PCR assays reveal the distribution of trichothecene genotypes of *F. graminearum* and *F. culmorum* isolates in Danish small grain cereals. *Int J Food Microbiol* 2012;157:384–392.

[72] Marin P, Jurado M, Magan N, Vázquez C, González-Jaén MT. Effect of solute stress and temperature on growth rate and *TRI5* gene expression using real time RT–PCR in *Fusarium graminearum* from Spanish wheat. *Int J Food Microbiol* 2010;140:169–174.

[73] Schmidt-Heydt M, Parra R, Geisen R, Magan N. Modelling the relationship between environmental factors, transcriptional genes and deoxynivalenol mycotoxin production by strains of two *Fusarium* species. *J R Soc Interface* 2011;8:117–126.

[74] Geisen R, Schmidt-Heydt M, Karolewiez A. A gene cluster of the ochratoxin A biosynthetic genes in *Penicillium*. *Mycotoxin Res* 2006;22:34–41.

[75] Schmidt-Heydt M, Baxter E Geisen R, Magan N. Physiological relationship between food preservatives, environmental factors, ochratoxin and *otapks*PV gene expression by *Penicillium verrucosum*. *Int J Food Microbiol* 2007;119:277–283.

[76] Schmidt-Heydt M., Graf E, Stoll D, Geisen R. The biosynthesis of ochratoxin A by *Penicillium* as one mechanism for adaptation to NaCl rich foods. *Food Microbiol* 2012;29:233–241.

[77] Geisen R. Molecular monitoring of environmental conditions influencing the induction of ochratoxin A biosynthesis genes in *Penicillium nordicum*. *Mol Nutr Food Res* 2004;48:532–540.

[78] O'Callaghan J, Dobson ADW. Molecular characterization of ochratoxin A biosynthesis and producing fungi. *Adv Appl Microbiol* 2005;58:227–243.

[79] Bacha N, Atoui A, Mathieu F, Liboz T, Lebrihi A. *Aspergillus westerdijkiae* polyketide synthase gene *"aoks1"* is involved in the biosynthesis of ochratoxin A. *Fungal Genet Biol* 2009;46:77–84.

[80] Gallo A, Knox BP, Bruno KS, Solfrizzo M, Baker SE, Perrone G. Identification and characterization of the polyketide synthase involved in ochratoxin A biosynthesis in *Aspergillus carbonarius*. *Int J Food Microbiol 2014*;179:10–17.

[81] Crespo-Sempere A, Marín S, Sanchis V, Ramos AJ. VeA and LaeA transcriptional factors regulate ochratoxin A biosynthesis in *Aspergillus carbonarius*. *Int J Food Microbiol* 2013;166:479–486.

[82] Tannous J, El Khoury R, Snini S. et al. Sequencing, physical organization and kinetic expression of the patulin biosynthetic gene cluster from *Penicillium expansum*. *Int J Food Microbiol* 2014;189:51–60.

[83] Puel O, Tadrist S, Delaforge M, Oswald IP, Lebrihi A. The inability of *Byssochlamys fulva* to produce patulin is related to absence of 6-methylsalicylic acid synthase and isoepoxydon dehydrogenase genes. *Int J Food Microbiol* 2007;115:131–139.

Antonio Moretti and Antonio F. Logrieco

5 Climate change effects on the biodiversity of mycotoxigenic fungi and their mycotoxins in preharvest conditions in Europe

5.1 Introduction

Changes in climate have already been apparent in temperatures and a warming of climate faster than the average global warming measured in the last century. Reliable reports from specialized institutions of the United Nations Intergovernmental Panel on Climate Change (IPCC) and the World Meteorological Organization (WMO) have confirmed that climate changes are occurring nowadays which may be connected with the ever-increasing human activities of a tremendously growing world population since the commencement of the industrial revolution. Meteorological extremities such as droughts, or even torrential rains, floods, and increasing frequency and duration of heat waves are connected to climate change. It is expected that such phenomena will increase during the coming decades.

Climate change has a direct and significant effect on agricultural production, diminishing thereby food security as well as creating public health risks, including decreasing safety of our foods [1]. Recently, particular emphasis has been placed on the climate shift regarding Europe. Challenges of climate change for European agriculture include the effects on crop yields and production risks such as heat waves, droughts, and pests. The global warming tendency [2] and increasing contamination and stress effects in relation to meteorological extremities, food and feed safety, as well as that of water sources may increase infections and poisoning occurrences in this region [3]. Drought stresses reduce phytoimmunity of crop plants, and extreme precipitation and heat waves increase the opportunity for plant pathogenic microorganisms to grow [4]. In particular, the EU green paper on climate change in Europe also suggests that effects will be regional and either detrimental or advantageous depending on the geographical area [5]. Thus, in southern Europe, changes may lead to a temperature increase of 4–5 °C with longer drought periods, resulting in increasing desertification and decreasing crop yields. In central Europe, the predictions are for an increase of 3–4 °C, and higher rainfall and floods, while in northern Europe a mean temperature increase of 3–4.5 °C and a significant increase in precipitation of 30–40 % are expected. This may lead to increases in crop yields and perhaps new crop cultivation patterns for central and northern Europe on the one hand, but could also dramatically affect the species profile and biochemical characterization of microbial communities occurring on crops with the introduction of new diseases [5].

Mycotoxins are low molecular weight toxic compounds produced by fungi which pose a serious risk to human and animal health worldwide. Cereals are often the most severely affected crops, and because they are a staple food for a large portion of humanity, mycotoxins are the most prevalent food-related health risk in field crops. Worldwide scientific and economic focus on mycotoxins also results from the significant economic losses associated with their negative impact on human health, animal productivity, and international trade.

The mycotoxins of greatest concern to food and feed safety are produced primarily by three genera of filamentous fungi: *Aspergillus*, *Fusarium*, and *Penicillium* [6]. In addition, although fungi can collectively produce hundreds of mycotoxins, only the following are of serious concern worldwide: aflatoxins, fumonisins, ochratoxins, patulin, trichothecenes, and zearalenone [6]. Some of these mycotoxins exhibit a very high degree of structural diversity, reflecting genetic diversity of the species they are produced by. Some of the mycotoxigenic species of the greatest concern occur worldwide. However, other important species have more limited distribution. In some cases mycotoxin contamination can result from the presence of multiple fungal species in a single crop.

An important component of efforts to control mycotoxin contamination problems is the study of the morphological, molecular genetic, metabolic, and plant pathological diversity of mycotoxigenic fungi. The plasticity of agriculturally relevant traits contributes to the collective ability of fungi to colonize a wide variety of crop species and to adapt to a range of environmental conditions. Knowledge of environmental factors which affect the ability of fungi to grow, survive, and interact with plants is important in order to better understand the variation in the population structures of mycotoxigenic fungi, their interactions with crop plants, and their ability to produce mycotoxins. Because climate can profoundly affect growth, distribution, and mycotoxin production in fungi, climate change has the potential to increase the risks that mycotoxigenic fungi pose to food and feed safety. The appearance of new mycotoxin-commodity combinations is of further concern and provides evidence for the emergence of new fungal genotypes with higher levels of aggressiveness and altered mycotoxin production. Trans-global trade of plant products can also contribute to the spread of toxigenic fungi and has led to increasing interest in the molecular diversity of toxigenic fungi on a global scale [7].

Fungal biodiversity is one of the most important contributors to the occurrence and severity of mycotoxin contamination of crop plants. Phenotypic and metabolic plasticity has enabled mycotoxigenic fungi to colonize a broad range of agriculturally important crops, especially cereals, and to adapt to a range of environmental conditions. New mycotoxin-commodity combinations provide evidence for the ability of fungi to adapt to changing conditions and the emergence of genotypes which confer enhanced aggressiveness toward plants and/or altered mycotoxin production profiles.

Beside the shifts in the profiles of mycotoxigenic genera and species which colonize important cereals, perhaps the most important contributor to qualitative differences in mycotoxin production among fungi is the variation in mycotoxin biosynthetic genes. Molecular genetic and biochemical analyses of toxigenic fungi have elucidated specific differences in biosynthetic genes responsible for intra- and interspecific differences in mycotoxin production. For *Aspergillus* and *Fusarium*, the mycotoxigenic genera of greatest concern, variations in biosynthetic genes responsible for production of individual families of mycotoxins appear to be the result of evolutionary adaptation. Examples of such variation have been reported for: (a) aflatoxin biosynthetic genes in *Aspergillus flavus*; (b) trichothecene biosynthetic genes within and among *Fusarium* species; and (c) fumonisin biosynthetic genes in *Aspergillus* and *Fusarium* species. Understanding variation in these biosynthetic genes and the basis for the variation in mycotoxin production is important for accurate assessment of the risks that fungi pose to food safety and for the prevention of mycotoxin contamination of crops in the field and in storage. On the other hand, it is also important to evaluate the main shifts of the toxigenic fungi colonizing cereals due to the impact of climate change in order to provide more accurate estimations of the mycotoxigenic risks related to these crops. The overview provided in this chapter will mainly refer to cereals since they are a staple food and extremely sensitive to mycotoxin contamination [8]. In particular aflatoxin and fumonisin contamination of maize and deoxynivalenol (DON) of minor cereals due to Fusarium head blight (FHB) will be considered in order to estimate future changes caused by both climatic factors and cultivation practices.

5.2 Climate change and the risk of aflatoxin and *Aspergillus* contamination in Europe

Aflatoxins are potent carcinogens which include four major structural analogues: AFB_1, AFB_2, AFG_1, and AFG_2. The International Agency for Research on Cancer (IARC) has classified AFB_1 as a group 1 carcinogen, i.e. carcinogenic to humans [9]. In addition to hepatocellular carcinoma, aflatoxins are associated with occasional outbreaks of acute aflatoxicosis which lead to death shortly after exposure [10]. Aflatoxins are produced in diverse agricultural products by several species of *Aspergillus*, but the two species of greatest concern are *A. flavus* and *A. parasiticus* [11].

Aflatoxin outbreaks are most severe in tropical and subtropical areas around the world, with temperate regions, such as the United States Midwest, also subject to aflatoxin contamination. Until 2004, the European perspective regarding aflatoxin contamination was confined to imported foods such as peanut cake, palm kernel, copra, and corn gluten meal (depending on origin; European Food Safety Authority (EFSA) [12]). Several surveys which have been conducted to detect AFs in feed samples in Europe found a small percentage of materials contaminated with AFB_1 concentrations above the regulatory limit. In fact, a survey of 110 maize samples in northern Italy in 2003, initially planned to monitor the occurrence of fumonisins, found 75 % positive AFB samples with a mean concentration of 4.4 and a maximum of 154.5 μg/kg [13].

In 2006, aflatoxin contaminated rice meal used in dairy cattle feed production was identified as the cause of elevated AFM_1 levels in Swedish milk.

However, a big survey conducted by the EFSA [14] established the emerging issue of potential aflatoxin contamination of corn, almonds, and pistachios grown in areas of southern Europe due to the subtropical climate which had occurred in recent years.

Changes in climate have already become apparent in temperatures and warming of the climate faster than the average global warming measured in the last century. Challenges of climate change for European agriculture include the effects on crop yields and production risks such as heat waves, droughts and pests. Due to the so-called "mediterraneanization" of southern Europe, aflatoxigenic species, formerly known in tropical and subtropical regions, may invade this region, which was of temperate climate in previous centuries. A shift in traditional occurrence areas for aflatoxins is therefore to be expected due to the increasing average temperatures. In this respect, the Mediterranean zones have been identified as a climate change hotspot where extreme changes in temperature, CO_2 levels, and rainfall patterns are predicted. Regarding aflatoxins, their contamination events are more prevalent during times of high heat and drought, which may stress the host plant and thereby facilitate *A. flavus* infection [15, 16].

In 2003/2004 and 2012/2013, hot and dry seasons led to severe *A. flavus* infection of maize in northern Italy. As a result of the very dry conditions in those years, *A. flavus* became a significant problem as a dominant pathogen in maize. The EFSA therefore issued a call at the end of 2009 for a project to investigate the possibility of the emergence of aflatoxin B1 in cereals in the EU due to climate change by modelling, predicting, and mapping. The results of this project [17] led to the main conclusion, based on the predictive model developed for *A. flavus* growth and AFB1 production linked to crop phenology data, that the risk of aflatoxin contamination will increase in maize in the future due to the climate change trend. In the +2 °C climate change scenario there is a clear increase in aflatoxin risk in areas such as central and southern Spain, the south of Italy, Greece, northern and southeastern Portugal, Bulgaria, Albania, Cyprus, and European Turkey as compared to the actual current temperature. In addition to the high aflatoxin risk in these southern European countries, low and medium aflatoxin risk at harvest in the four main maize producing countries (Romania, France, Hungary, and northeast Italy) were predicted. It will be therefore be more important across Europe each year to: (1) improve harmonization of surveillance and monitoring of aflatoxins; (2) improve databases on the geographical distribution and prevention methods for aflatoxins; (3) develop models for the prediction of aflatoxin contamination in the new biogeographical agricultural scenarios.

The ability to produce aflatoxins is highly conserved in some species but variable in others. For example, 94–97 % of the *A. parasiticus* strains which have been examined produce aflatoxins, whereas production in *A. flavus* is highly variable and depends on genotype, substrate, and geographic origin [18, 19].

The aflatoxin biosynthetic gene (*afl*) cluster includes 25 genes [20]. The gene content and organization of the cluster is highly conserved among *Aspergillus* species in section *Flavi*, which includes *A. flavus* and *A. parasiticus*. Sequence variability and deletions in various genes/regions of the *afl* cluster have also been used to assess variability in *A. flavus* [21]. Moreover, differences in *afl* genes have been used to distinguish between aflatoxin-producing and nonproducing strains of *A. flavus* and *A. parasiticus*. Understanding such genetic differences is important because aflatoxin-nonproducing strains of *A. flavus* are used to control aflatoxin contamination in some crops [22]. Recently, Gallo et al. [23] examined a collection of aflatoxin-producing and nonproducing isolates of *A. flavus* for the presence of seven *afl* genes, two regulatory genes *aflR* and *aflS*, and the structural genes *aflD*, *aflM*, *aflO*, *aflP*, and *aflQ*. The result was the grouping of strains into four different amplification patterns. All aflatoxin-producing isolates yielded the complete set of amplification products, whereas nonproducing isolates did not yield products for three, four, or all seven genes, indicating a high level of genetic variability among *A. flavus* isolates. Together, analyses of variation in the *afl* cluster in *A. flavus* have illustrated the genetic complexity of this species, which includes variability in the *afl* cluster. This genetic variability has provided markers which can be used to monitor variation in *A. flavus* and to evaluate the risk they pose when present on food commodities.

The genetic diversity of *A. flavus* populations collected from maize kernels in northern Italy from 2003 to 2010 was also assessed by Mauro et al. [24], who evaluated the presence or absence of several aflatoxin genes. Six deletion patterns of genes in aflatoxin clusters were detected. Regarding the atoxigenic isolates, some had no deletion in the cluster, others had the entire cluster deleted, and only a single strain had a deletion pattern with only two genes amplified of the thirteen tested.

Therefore, the genetic variability of aflatoxin cluster in non-aflatoxigenic isolates appears diverse and complex but its understanding is important for the selection of safe and effective nonproducing strains which could potentially be used for biocontrol to limit aflatoxin contamination.

Magan et al. [25] have shown that in addition to temperature, changes in CO_2 concentration and water activity can influence the growth and mycotoxin production of some mycotoxigenic species, including *A. flavus*, especially under water stress. Moreover, Medina et al. [26] also studied the effects of the interaction of a_w, temperature, and elevated CO_2 on ten structural and regulatory aflatoxin biosynthetic gene expression. The data generated showed that these interactions have a significant impact on gene expression stronger than on the growth of *A. flavus* and can stimulate the AFB_1 production. Therefore, due to the effect that the above mentioned environmental factors have on the expression of *A. flavus* aflatoxin genes, accurate knowledge of the genetic variability of the *A. flavus* populations occurring in the field is of extreme importance for both phylogenetic relationships among fungal strains and mycotoxin gene biosynthetic pathways. This would strengthen the possibility of predicting aflatoxin production according to changes in environmental factors and would allow better management of aflatoxin risk in the field.

In conclusion, the occurrence of AFB_1 at high levels in Europe in the years 2003–2004 and 2012–2013 underlines the fact that climate change will entail a change in the mycotoxin distribution patterns observed today. Global trade of plant products can also contribute to the spread of aflatoxigenic fungi and to the increase of diversity of local fungal populations. The study of the biodiversity of aflatoxigenic fungi occurring in maize in Europe is essential for the development of strategies for the control of aflatoxin contamination. In this regard, the molecular characterization of native atoxigenic strains, acting through competitive exclusion of aflatoxin producers, with superior adaptation to a geographical region, should provide benefit of long-term displacement of toxigenic strains in maize environment. Finally, the development of predictive models for aflatoxin occurrence based on regional weather data would be a valuable tool for estimating the risk of contamination after a given growing season, together with the use of biopesticides in the framework of integrated pest management.

5.3 Fusarium head blight (FHB) of cereals: impact of climate change on the risk of trichothecenes and *Fusarium* contamination in Europe

The fungal genus *Fusarium* consists of over 90 described species and likely many additional as yet undescribed but phylogenetically distinct species [27–30]. Many of these species are plant pathogens and produce a range of mycotoxins, the most agriculturally important being trichothecenes and fumonisins [31]. Trichothecenes are a family of terpene-derived mycotoxins produced by multiple species of *Fusarium* and are among the most economically significant mycotoxins worldwide because of their potency and widespread occurrence in cereals [32]. These mycotoxins are strong inhibitors of eukaryotic protein biosynthesis, causing a wide range of toxic effects in humans and animals, including vomiting, hemorrhagic, and immunosuppressive effects. They can be toxic to plants and contribute to the pathogenesis of *Fusarium* on some crops [31]. All trichothecenes have a tricyclic skeleton structure with an epoxide group, but they can be divided into two structurally distinct groups based on the absence (type A trichothecenes) and presence (type B trichothecenes) of a keto group at carbon atom 8 (C-8) of the skeleton. The type A trichothecenes of greatest concern are T-2 toxin and HT-2 toxin, while the type B trichothecenes of greatest concern are DON, nivalenol (NIV), and their acetyl-derivatives [31].

5.3.1 Organization of *TRI* loci and trichothecene structural variation

Multiple studies have revealed that structural variation of trichothecene mycotoxins produced by different species of *Fusarium*, and in some cases different strains of the same species, result from DNA sequence polymorphism causing variation in the functionality of trichothecene gene cluster (*TRI* genes). The different organization of the *TRI* genes among the genomes of the *Fusarium* species shows how widely genetic diversity can also be expressed in the biosynthetic gene pathways along this important phytopathogenic and toxigenic fungal genus.

Fusarium graminearum sensu stricto (*Fusarium graminearum s.s.*) and *F. sporotrichioides* are the two species in which the trichothecene biosynthetic gene cluster was first characterized. In both species, the cluster consists of 12 genes which are responsible for synthesis of the core trichothecene skeleton and several modifications to it. Both species also have two smaller *TRI* loci. The first of these consists of one gene, *TRI101*, and the second locus consists of two genes, *TRI1*, and *TRI16* [33]. Analysis of *TRI* loci in 16 species of *Fusarium* revealed that *TRI1* and *TRI101* are located in the core *TRI* cluster in four species of *Fusarium* [33] which are members of the *F. incarnatum-equiseti* species complex [28]. Relocation of *TRI1* and *TRI101* into the core *TRI* cluster of the *F. incarnatum-equiseti* species complex provided evidence for growth of a fungal secondary metabolite gene cluster by gene relocation rather than by gene duplication and subsequent divergence of pre-existing cluster genes [33]. *TRI1* was shown to be a polymorphic gene. This polymorphism contributes to the differences in function of *TRI1* which are responsible for the difference in production of type B or type A trichothecenes by *F. graminearum* and *F. sporotrichioides*, respectively [34]. Moreover, polymorphism in *TRI13* is responsible for variations in the presence and absence of a C-4 hydroxyl, an important structural difference which occurs within type B trichothecenes. Strains which differ according to their ability to produce different types of trichothecenes are defined as chemotypes.

Similar patterns of deletions within *TRI13* occur in isolates of the so-called *F. graminearum* clade (*F. graminearum* species complex (FGSC), *F. crookwellense*, and *F. culmorum* [31]). Strains of these fungi with a functional *TRI13* produce the C-4 hydroxylated trichothecene NIV, whereas strains with a nonfunctional *TRI13* (i.e. with the deletions) produce DON [31]. Another trichothecene chemotype difference exhibited by members of the *F. graminearum* clade is production of 3-acetyldeoxynivalenol (3-ADON) versus 15-acetyldeoxynivalenol (15-ADON). The genetic basis for this difference resides in the *TRI* cluster gene *TRI8*. Alexander et al. [35] identified consistent DNA sequence differences in the coding region of *TRI8* in 3-ADON and 15-ADON strains. They also showed that *TRI8* in NIV-producing *Fusarium* strains functions like that in 15-ADON strains, and that *TRI3* was functional in all three chemotypes. Together, this data indicated that differential activity of *TRI8* determines the 3-ADON and 15-ADON chemotypes in *Fusarium*.

5.3.2 FHB of minor cereals

FHB, caused by a complex of *Fusarium* species, is one of the most important fungal diseases associated with wheat and several other minor cereals, the main species of which is *F. graminearum s.s.* The effects of FHB are related to yield and quality reduction of the infected kernels, due to the accumulation of mycotoxins, especially DON, in the raw grain and in the processed wheat products [36]. *Fusarium graminearum* is more appropriately defined as FGSC, since in the last decade it has been described as a group of related species, differing in geographical distribution, mycotoxin production, and pathogenicity [30]. According to the latest reports, FGSC consists of 15 different species based on multilocus phylogenetic analyses [30]. The species *F. graminearum s.s.* is the most common and widespread species of the FGSC, prevalent in North America [37] and Europe [38], but reported worldwide, whereas other species are geographically more restricted. As an example, *F. asiaticum* is a widely distributed species in Asia and South America [39], *F. gerlachii* and *F. louisianense* are typical of US [7, 30], *F. aethiopicum* is endemic to Africa [40], *F. acaciae-mearnsii* to Australia and South Africa [39, 41], *F. nepalense* is a new species recently discovered in Nepal [30], *F. vorosii* and *F. ussurianum* are endemic to Asia [7, 42], *F. boothi* and *F. mesoamericanum* are typical of Central America [39, 40], and *F. cortaderiae*, *F. brasilicum*, *F. austroamericanum* and *F. meridionale* are typical of South America [39]. However, strains of *F. meridionale* and *F. boothii* were also discovered in South Africa [41], in South Korea [43], and in Nepal [44], suggesting that world trade could contribute to the spread of some FGSC species to countries in which they are not endemic [42].

Besides *F. graminearum s.s.*, many other *Fusarium* species have been associated worldwide with FHB, such as *F. avenaceum*, *F. culmorum* and *F. poae* primarily, and *F. acuminatum*, *F. cerealis* (syn. *F. crookwellense*), *F. equiseti*, *F. langsethiae*, *F. sporotrichioides*, and *F. tricinctum* to a lesser extent. The toxigenic potential of FHB pathogens can greatly vary between species, since each can produce a specific range of mycotoxins [31]. Although agricultural practices play a key role in the prevalence of FHB pathogens, the predominance of FHB species is determined, to a greater extent, by climatic factors, particularly temperature and moisture [45], and can therefore dramatically change in different geographical areas. In particular Europe, which is characterized by a wide range of climatic conditions from the southern (Mediterranean countries) to central and northern regions, shows dramatic differences in the species composition associated with FHB [8]. Indeed, these pathogens can occur on wheat heads in different combinations leading to multiple mycotoxin contaminations in the harvested grain. It is therefore of key importance to determine the exact nature of the FHB complex for more reliable disease prediction and management, and in order to manage the food risks associated with consumption of contaminated kernels.

F. graminearum is the most common cereal head blight pathogen in moist, warm continental climates of Europe, while *F. culmorum* and *F. avenaceum* are more prevalent in maritime and cooler European regions. Inoculum production of the *Fusarium* species is dependent on rainfall, and warm and moist conditions during anthesis are key factors for disease development [46]. The species differ in their climatic distribution and in the optimum climatic conditions they require [47]. Significant yield losses and mycotoxin accumulation have been reported under the hot, wet climatic conditions favorable for *F. graminearum*, while the species is sensitive to lower temperatures [48]. Disease progress can cease at cool temperatures (lower than 15 °C) and a significantly longer wet period is required at temperatures of 20 °C as opposed to 25 °C, while perithecial production and development is optimal at temperatures between 15 °C and 24 °C and limited in higher or lower temperatures and dry conditions [49]. Cool and moist conditions at the end of summer season favor *F. avenaceum* infection because it can infect at lower temperatures when there is enough humidity. On the other hand, *F. poae* favor a warm dry environment, and precipitation has a negative effect on infection, while correlations between temperatures over 15 °C at maturity and *F. poae* infection were positive. *F. tricinctum* favors precipitation after, but not during flowering. *F. culmorum* prefers cooler temperatures than *F. graminearum*, while higher humidity is also more favorable to *F. graminearum* than *F. culmorum*. Langseth and Elen [50] observed that spring drought seems to make cereals more susceptible to *Fusarium* infection and high DON concentrations. *F. langsethiae* infects flowers and developing kernels early in the season, when high humidity occurs [51], but it can also be prevalent on oats in dry conditions [52]. Finally, exposure to cold and wet weather due to late harvest can also lead to high DON contamination [50].

The cool and rather humid climate in northern Europe influences the *Fusarium* flora. The dominant species reported in Finland is *F. avenaceum*, together with the related species *F. arthrosporioides*, which is also the most common *Fusarium* species in all Scandinavian countries [53]. This species does not produce trichothecene mycotoxins, but minor toxic metabolites such as moniliformin, beauvericin, and enniatins [31]. *Fusarium tricinctum* is also relatively common on minor cereals depending on the year, and produces enniatins and moniliformin [54]. The incidence of *F. poae*, infection of which is favored by dry weather readily at ear emergence, has been recently observed to have increased in Norway [55]. Moreover, *F. poae* is not considered a seedling pathogen, does not affect plant development, does not cause head blight symptoms [56], and produces beauvericin, and both types A and B trichothecene [31]. While *F. graminearum* is the main DON producer in central and southern Europe, in Nordic areas *F. culmorum* has been reported as dominant [8], being common on cereal roots, stem base, and heads in all Scandinavian countries, including Finland [8]. *Fusarium culmorum* tolerates cold conditions, survives very well in crop cereal debris and soil, its infection is induced by high temperatures and rainfall during anthesis, and can proceed until harvest. However, a decline in the presence of *F. culmorum* and an increase in *F. graminearum* have been reported in the last decade in some areas of central and northern Europe by several authors [57, 58].

Fusarium langsethiae, a species formally described in the last decade, is very common in all northern European countries, often being the first colonizer of kernels after ear emergence [51]. *Fusarium langsethiae* produces T-2 and HT-2, the most harmful type A trichothecene toxins [31] also in cool and humid conditions. According to Pettersson et al. [59], T-2/HT-2 contaminations have increased on minor cereals during the past 20 years in northern Europe due to increased infections of *F. langsethiae*.

In the cooler, maritime climate of Britain and the Netherlands, the most common species involved in FHB in cereal grain are *F. culmorum*, *F. graminearum s.s.*, *F. avenaceum*, and *F. poae*. At the beginning of 2000, *F. graminearum* was the most abundant *Fusarium* species on wheat in the Netherlands; a significant change in the situation had occurred compared with the early 1990s when *F. culmorum* was the dominating species in wheat [57]. The same was observed in the UK, where *F. graminearum* has increased on wheat, while the previously common *F. culmorum* has become less important [60]. However, also in the cooler temperate climates of Europe areas such as Germany, where *F. culmorum* used to be the prevalent species, *F. graminearum* has become the dominant species in the last decade, because the higher temperatures favor its dominance in the FHB complex [61]. In more eastern locations, such as Poland, *F. poae* has been the predominant species for a long period, followed by *F. tricinctum*, *F. avenaceum*, *F. culmorum*, and *F. graminearum* [8]. However, a significant increase in the frequency of *F. graminearum* was observed during the last decade in all regions of Poland, including the northern areas, as reported by Stępień and Chełkowski [62].

T-2 and HT-2 toxin contamination have become more prevalent on oats and barley in the UK, which has been connected to *F. langsethiae* detection in grains [60]. In addition, in northern France, this species has become prevalent on barley in recent years [63]. The species grows and produces toxins in a wide temperature range [25]. *Fusarium langsethiae* is an emerging mycotoxin producer also in Poland, where Lukanowski and Sadowski [64] reported the occurrence of infection on winter wheat.

The species profile of strains causing FHB in southern Europe varies annually depending on environmental conditions. In general, FHB incidence is low and in the most southern regions of Italy and Spain the disease is absent. However, in the more northern regions of Italy, Spain, and Portugal, southern France, and the whole Balkan Peninsula, *F. graminearum* and *F. poae* are reported to be the largely dominant species on cereals at maturity, with DON being the most frequently occurring mycotoxin on grains and *F. avenaceum* and *F. culmorum* rarely associated with FHB diseased plants [8]. However, more recently, in the southern regions of the Balkan Peninsula, *F. verticillioides*, and, to a lesser extent, *F. sporotrichioides* have been more frequently identified on cereals, showing fumonisins [65] and T-2 and HT-2 [66], respectively, as possible further mycotoxin risk associated with *Fusarium* colonization of cereal kernels. Interestingly, increased occurrence of *F. langsethiae* has been reported in Italy, with an increase of the T-2 and HT-2 contamination of kernels [67].

Trichothecene production by *Fusarium* species is broadly species specific [8]. The most important mycotoxins produced by FGSC members are type B trichotecenes, which are mainly produced by *F. graminearum s.s.*, the most frequent and virulent species within the FHB complex, usually occurring in warm and humid environmental conditions [38, 45]. Three chemotypes can be distinguished within FGSC populations: the NIV chemotype for strains producing NIV and 4NIV; the 3ADON for strains producing DON and 3ADON; and the 15ADON chemotype for strains producing DON and 15ADON. Some strains produce both DON and NIV and have been defined as "unknown chemotype" [68]. The DON chemotypes are pathogenically more aggressive towards wheat, compared to the NIV chemotype, probably reflecting a selective advantage [69]. Various investigations worldwide showed that DON chemotypes (in particular the 15ADON chemotype) predominate in North America, South America and Europe, while the NIV chemotype is more common in Europe than in North or South America. More recent surveys indicate that in northwestern Europe *F. graminearum sensu lato* is segregating for all three chemotypes [70, 71]. An increasing frequency of 3ADON was observed in North America besides the prevalent 15ADON [68, 72]. In addition, the NIV chemotypes are being replaced by DON-producing strains in Russia [42]. In contrast the presence of all DON and NIV chemotypes was reported in China, Japan, Korea and Nepal [73–75].

5.3.3 Impact of climate change on the Fusarium species profile associated with FHB

Several reports have tried to comprehend the possible shifting of *Fusarium* species involved in FHB in different geographic areas according to future environmental scenarios. According to Parikka et al. [53], the North European climate is predicted to become milder and more humid towards 2050. The subsequent climate change effects will therefore lead to a slight shift in composition of *Fusarium* flora in cereal grains [53]. *Fusarium poae* may become more prevalent in dry years, while the predicted increase of spring drought would benefit *F. culmorum*, which would otherwise be decreasing with higher temperatures [53]. Moreover, *F. langsethiae* will possibly utilize dry spring and summer periods and may infect the expanding winter wheat crops to some extent, in addition to oats and barley where it is now mainly detected [53]. Finally, the abovementioned authors suggested that tillage practices need to be changed towards conservation and reduced tillage, which predispose cereals to head blight infections [53].

Madgwick et al. [76] studied how the impact of climate change on wheat anthesis date could influence the impact of FHB in UK mainland arable areas. The authors used a wheat growth model for projections of incidence of anthesis date, and a weather-based model was developed for use in projections of incidence of FHB in the UK. Daily weather data, generated for 14 sites in arable areas of the UK for a baseline scenario and for high and low CO_2 emissions in the 2020s and 2050s, were used to project wheat anthesis dates and FHB incidence for each site and climate change scenario. The incidence of FHB was related to rainfall during anthesis and temperature during the preceding 6 weeks. The conclusion of this study was that, with the expected climate change in 2050, wheat anthesis dates will be earlier and FHB epidemics will be more severe, especially in southern England, mainly because of an increase of *F. graminer-aum* and the associated DON.

A relevant study outside Europe was carried out by Zhang et al. [77] and investigated the impacts on FHB on wheat in China due to climate change. The estimation of incidence of wheat FHB in central China was evaluated by developing a logistic weather-based regression model, using up to 10 years (2001–2010) of disease, anthesis date, and weather data available for 10 locations in Anhui and Hubei provinces. How climate change may affect wheat anthesis date and FHB in central China was estimated for the period 2020–2050 using the wheat growth model Sirius and simulated weather data obtained employing the regional climate modelling system PRECIS [77]. The study suggested that climate change will increase the risk of serious FHB epidemics on winter wheat in central China by the middle of this century, as also predicted for the UK by Madgwick et al. [76]. The implications for food security in China are extremely serious, because FHB is already a cause of substantial losses in wheat yield and quality, exacerbated by the high level of *Fusarium* mycotoxin accumulation in the kernels [77]. Because wheat is an essential staple crop over large areas of China, these predicted impacts of climate change would provide a guide to the Chinese government and industry in preparation for adaptation to climate change [77].

Other weather-based models for predicting the risk of FHB have been developed for Argentina, the USA, Italy, Brazil, and the UK [76, 78–81]. All these models develop country specific projections of impacts of climate change. However, they also need to incorporate impacts of climate change on crops by combining a crop growth model as performed by Madgwick et al. [76]. This latter model evaluated the impact of climate change on important crop stages such as date of anthesis, which is a key factor for FHB development, because *Fusarium* species infect during anthesis [82]. The model used by Zhang et al. [77] showed that incidence of FHB is related to the number of days of rainfall in a 30-day period after anthesis, and that high temperatures 2 to 3 weeks before anthesis increase the incidence of disease. The relationship with rainfall after anthesis is related to the effect of rain on infection and subsequent growth of *Fusarium* within the wheat ear [82], while high temperatures before anthesis may relate to the influence of increased temperature on *Fusarium* growth and sporulation [47].

Besides the environmental conditions, further important factors affect FHB epidemic development, such as agronomic practices. This limits the efficacy of weather-based disease models [77]. In particular, the proximity of previous maize crops can dramatically affect FHB development, because maize debris is a potent source of inocula of *F. graminearum* [83]. An expected increase of maize in rotations can be considered a significant risk factor for *F. graminearum* infections, especially in no-till practices, since in warm and humid conditions the ascospore production in crop residues could cause frequent epidemics. Therefore, an indirect impact of climate change on the severity of FEB epidemics will be the increase of maize cultivation in many geographic areas where the future climate scenario will allow farmers to grow this crop, well adapted to the higher temperatures [73].

In general, the information obtained from all investigations on the impacts of climate change on FHB will allow the definition of strategies for both governments and industry for reduction of the risk related to the more favorable conditions for FHB and mycotoxin contamination of grains. New wheat cultivars with higher tolerance and more effective fungicide treatments to control *Fusarium* species which cause FHB should be considered [84]. Since both strategies could require many years to become effective, it is important to begin adapting to the impacts of climate change on crop disease soon. Such strategies for adaptation to impacts of climate change on FHB will contribute to sustainable wheat production and improved food safety and security [73].

In conclusion, the impacts of climate change on FHB could result in a dramatic increase of disease severity worldwide with higher risks due to the mycotoxin contamination of the end products. Moreover, the deep profile modifications of toxigenic *Fusarium* species occurring on kernels at maturity in different geographical areas of the world will cause the development of new mycotoxin risks in specific regions, due to the changed ability of given *Fusarium* species to colonize new environments. Examples are the increase of DON in the most northern regions of Europe, the higher risk of contamination by T-2 and TH-2 toxins in many regions of Europe, and the increase of fumonisin risk in the Balkan Peninsula. Finally, the influence of climate change on the expression of mycotoxin biosynthetic genes in the different *Fusarium* species needs to be more thoroughly investigated. This knowledge will be of key importance in understanding the genetic mechanisms used by the toxigenic *Fusarium* species to increase genetic variability and will allow better definition of strategic solutions for disease management.

References

[1] WHO. *Protecting health in Europe from climate change*. Menne B, Apfel F, Kovats S, Racioppi F eds. Copenhagen, Denmark: World Health Organization Regional Office for Europe; 2008.

[2] Nature Editorial on Berkeley Earth Surface Temperature (BEST) studies. Scientific climate. *Nature* 2011;478:428.

[3] Farkas J, Beczner J. A klímaváltozás hatásai az élelmiszerbiztonságra (Potential effects of climate change on the safety of foods). *Élelmiszervizsgálati Közlemények* 2010;56:219–230.

[4] Tirado MC, Clarke R, Jaykus LA, McQuatters-Gollop A, Frank J. Climate change and food safety: A review. *Food Res Int* 2010;43:1745–1765.

[5] European Commission. *Adapting to climate change in Europe–options for EU action. Green paper from the Commission to the Council, the European Parliament, the European Economic and Social Committee of the Regions*. COM (2007) 354. Brussels, Belgium: European Commission: 2007. (Accessed December 5, 2014, at http://eur-lex.europa.eu/legal-content/EN/TXT/?uri=CELEX:52007DC0354).

[6] Marasas WFO, Gelderblom WCA, Shephard GS, Vismer HF. Mycotoxins: A global problem. In: Leslie JF, Bandyopadhyay R, Visconti A (eds). *Mycotoxins: Detection methods, management, public health and agricultural trade*. Wallingford, Oxfordshire, UK: CABI; 2008. p. 29–40.

[7] Starkey DE, Ward TJ, Aoki T et al. Global molecular surveillance reveals novel Fusarium head blight species and trichothecene toxin diversity. *Fungal Genet Biol* 2007;44:1191–1204.

[8] Logrieco A.F, Moretti A. Between emerging and historical problems: An overview of the main toxigenic fungi and mycotoxin concerns in Europe. In: Leslie JF, Bandyopadhyay R, Visconti A (eds). *Mycotoxins: Detection methods, management, public health and agricultural trade.* Wallingford, Oxfordshire, UK: CABI; 2008. p. 139–153.

[9] IARC. *Some naturally occurring substances: Food items and constituents, heterocyclic aromatic amines and mycotoxins. IARC Monographs on the evaluation of carcinogenic risks to humans, vol. 56.* UK: International Agency for Research on Cancer; 1993.

[10] Azziz-Baumgartner E, Lindblade K, Gieseker K et al. Case-control study of an acute aflatoxicosis outbreak, Kenya, 2004. *Environ Health Perspect* 2005;113:1779–1783.

[11] Frisvad JC, Skouboe P, Samson RA. Taxonomic comparison of three different groups of aflatoxin producers and a new efficient producer of aflatoxin B_1, sterigmatocystin and 3-*O*-methylsterigmatocystin, *Aspergillus rambellii* sp. nov. *Syst Appl Microbiol* 2005;28:442?453.

[12] European Food Safety Authority. Opinion of the Scientific Panel on Contaminants in the Food Chain on a request from the Commission related to aflatoxin B_1 as undesirable substance in animal feed. *EFSA J* 2004;39:1–27.

[13] Piva G, Battilani P, Pietri A. Emerging issues in southern Europe: Aflatoxins in Italy. In: Barug D, Bhatnagar D, van Egmond HP, van der Kamp J W, van Osenbruggen WA, Visconti A (eds). *The mycotoxin fact book: Food and feed topics.* Wageningen, the Netherlands: Academic Publishers; 2006. p. 139–153.

[14] European Food Safety Authority. Opinion of the Scientific Panel on Contaminants in the Food Chain on a request from the Commission related to the potential increase of consumer health risk by a possible increase of the existing maximum levels for aflatoxins in almonds, hazelnuts and pistachios and derived products. *EFSA J.* 2007;446:1–127.

[15] Schmidt-Heydt M, Abdel-Hadi A, Magan N, Geisen R. Complex regulation of the aflatoxin biosynthesis gene cluster of *Aspergilus flavus* in relation to various combinations of water activity and temperature. *Int J Food Microbiol* 2009;135:231–237.

[16] Mohale S, Medina A, Magan N. Effect of environmental factors on *in vitro* and *in situ* interactions between atoxigenic and toxigenic *A. flavus* strains and control of aflatoxin contamination of maize. *Biocontrol Sci Techn* 2013;23:776–793.

[17] Battilani P, Rossi V, Giorni P et al. *Modelling, predicting and mapping the emergence of aflatoxins in cereals in the EU due to climate change.* European Food Safety Authority (EFSA) supporting publication; 2012. (Accessed December 5, 2014, at http://www.efsa.europa.eu/en/supporting/pub/223e.htm)

[18] Vaamonde G, Patriarca A, Pinto VF, Comerio R, Degrossi C. Variability of aflatoxin and cyclopiazonic acid production by *Aspergillus* section *Flavi* from different substrates in Argentina. *Int J Food Microbiol* 2003;88:79–84.

[19] Pildain MB, Vaamonde G, Cabral D. Analysis of population structure of *Aspergillus flavus* from peanut based on vegetative compatibility, geographic origin, mycotoxin and sclerotia production. *Int J Food Microbiol* 2004;93:31?40.

[20] Yu J, Chang PK, Ehrlich KC et al. Clustered pathway genes in aflatoxin biosynthesis. *Appl Environ Microbiol* 2004;70:1253?1262.

[21] Chang PK, Ehrlich KC, Hua SST. Cladal relatedness among *Aspergillus oryzae* isolates and *Aspergillus flavus* S and L morphotype isolates. *Int J Food Microbiol* 2006;108:172?177.

[22] Dorner JW, Horn BW. Separate and combined applications of nontoxigenic *Aspergillus flavus* and *A. parasiticus* for biocontrol of aflatoxin in peanuts. *Mycopathologia* 2007;163:215?223.

[23] Gallo A, Stea G, Battilani P, Logrieco AF, Perrone G. Molecular characterization of an *Aspergillus flavus* population isolated from maize during the first outbreak of aflatoxin contamination in Italy. *Phytopathol Mediterr* 2012;51:198–206.

[24] Mauro A, Battilani P, Callicott KA, Giorni P, Pietri A. Cotty PJ. Structure of an *Aspergillus flavus* population from maize kernels in northern Italy. *Int J Food Microbiol* 2013;162:1–7.

[25] Magan N, Medina A, Aldred D. Possible climate change effects on mycotoxin contamination of food crops pre- and postharvest. *Plant Pathol* 2011;60:150–163.

[26] Medina A, Rodriguez A, Magan N. Effect of climate change on *Aspergillus flavus* and aflatoxin B_1 production. *Front Microbiol* 2014;5:348. doi: 10.3389/fmicb.2014.00348.

[27] Leslie JF, Summerell BA. *The* Fusarium *laboratory manual*. Ames, Iowa, USA: Blackwell Publishing; 2006.

[28] O'Donnell K, Sutton DA, Rinaldi MG, Gueidan C, Crous PW, Geiser DM. Novel multilocus sequence typing scheme reveals high genetic diversity of human pathogenic members of the *Fusarium incarnatum-F. equiseti* and *F. chlamydosporum* species complexes within the United States. *J Clin Microbiol* 2009;47:3851–3861.

[29] O'Donnell K, Rooney AP, Proctor RH et al. Phylogenetic analyses of *RPB1* and *RPB2* support a middle Cretaceous origin for a clade comprising all agriculturally and medically important fusaria. *Fungal Genet Biol* 2013;52:20–31.

[30] Sarver BA, Ward TJ, Gale LR et al. Novel Fusarium head blight pathogens from Nepal and Louisiana revealed by multilocus genealogical concordance. *Fungal Genet Biol* 2011;48:1077–1152.

[31] Desjardins AE. Fusarium *mycotoxins: Chemistry, genetics and biology*. St. Paul, Minnesota, USA: APS Press; 2006.

[32] CAST. *Mycotoxins: Risks in plant, animal and human systems, Task Force Report № 139*. Ames, Iowa, U.S.A: Council for Agricultural Science and Technology (CAST); January 2003.

[33] Proctor RH, McCormick SP, Alexander NJ, Desjardins AE. Evidence that a secondary metabolic biosynthetic gene cluster has grown by gene relocation during evolution of the filamentous fungus *Fusarium*. *Mol Microbiol* 2009;74:1128–1142.

[34] McCormick SP, Alexander NJ, Proctor RH. Heterologous expression of two trichothecene P450 genes in *Fusarium verticillioides*. *Can J Microbiol* 2006;52:220–226.

[35] Alexander NJ, McCormick P, Waalwijk C, van der Lee T, Proctor RH. The genetic basis for 3-ADON and 15-ADON trichothecene chemotypes in *Fusarium*. *Fungal Genet Biol* 2011;48:485–495.

[36] van der Fels-Klerx HJ, Stratakou I. T-2 toxin and HT-2 toxin in grain and grain-based commodities in Europe: occurrence, factors affecting occurrence, co-occurrence and toxicological effects. *World Mycotoxin J* 2010;3:349–367.

[37] Ward TJ, Clear RM, Rooney AP et al. An adaptive evolutionary shift in Fusarium head blight pathogen populations is driving the rapid spread of more toxigenic *Fusarium graminearum* in North America. *Fungal Genet Biol* 2008;45:473–484.

[38] Yli-Mattila T. Ecology and evolution of toxigenic *Fusarium* species in cereals in northern Europe and Asia. *J Plant Pathol* 2010;92:7–18.

[39] O'Donnell K, Ward TJ, Geiser DM, Kistler HC, Aoki T. Genealogical concordance between the mating type locus and seven other nuclear genes supports formal recognition of nine phylogenetically distinct species within the *Fusarium graminearum* clade. *Fungal Genet Biol* 2004;41:600–623.

[40] O'Donnell K, Ward TJ, Aberra D et al. Multilocus genotyping and molecular phylogenetics resolve a novel head blight pathogen within the *Fusarium graminearum* species complex from Ethiopia. *Fungal Genet Biol* 2008;45:1514–1522.

[41] Boutigny AL, Ward TJ, Van Coller GJ et al. Analysis of the *Fusarium graminearum* species complex from wheat, barley, and maize in South Africa provides evidence of species-specific differences in host preference. *Fungal Genet Biol* 2011;48:914–920.

[42] Yli-Mattila T, Gagkaeva T, Ward TJ, Aoki T, Kistler HC, O'Donnell K. A novel Asian clade within the *Fusarium graminearum* species complex includes a newly discovered cereal head blight pathogen from the Russian Far East. *Mycologia* 2009;101:841–852.

[43] Lee YW, Jeon JJ, Kim H et al. Lineage composition and trichothecene production of *Gibberella zeae* population in Korea. In: Yoshizawa T (ed). *New horizons of mycotoxicology for assuring food safety.* Kagawa, Japan: Japanese Association of Mcyotoxicology; 2004. p. 117–122.

[44] Desjardins AE, Proctor RH. Genetic diversity and trichothecene chemotypes of the *Fusarium graminearum* clade isolated from maize in Nepal and identification of a putative new lineage. *Fungal Biol* 2011;115:38–48.

[45] Xu XM, Nicholson P, Thomsett et al. Relationship between the fungal complex causing Fusarium head blight of wheat and environmental conditions. *Phytopathology* 2008;98:69–78.

[46] Xu X. Effects of environmental conditions on the development of *Fusarium* ear blight. *Eur J Plant Pathol* 2003;109:683–689.

[47] Doohan FM, Brennan J, Cooke BM. Influence of climatic factors on *Fusarium* species pathogenic to cereals. *Eur J Plant Pathol* 2003;109:755–768.

[48] Parry DW, Jenkinson P, McLeod L. *Fusarium* ear blight (scab) in small grain cereals – a review. *Plant Pathol*. 1995;44:207–238.

[49] Dufault NS, De Wolf ED, Lipps PE, Madden LV. Role of temperature and moisture in the production and maturation of *Gibberella zeae* perithecia. *Plant Dis* 2006;90:637–644.

[50] Langseth W, Elen O. Differences between barley, oats and wheat in the occurrence of deoxynivalenol and other trichothecenes in Norwegian grain. *J Phytopathol* 1996;144:113–118.

[51] Parikka P, Hietaniemi V., Rämö S. The effect of tillage on *Fusarium* infection and mycotoxins on barley and oats. In: *The BCPC International Congress Crop Science & Technology 2005: Congress Proceedings, vol. 1, SECC, 31 Oct-2 Nov 2005. Glasgow, UK.* British Crop Protection Council; 2005. p. 423–428.

[52] Parikka P, Hietaniemi V, Rämö S, Jalli H. *Fusarium* infection and mycotoxin contents of oats under different tillage treatments. *J Plant Pathol* 2008;90(S3):75.

[53] Parikka P, Hakala K, Tiilikkala K. Expected shifts in *Fusarium* species' composition on cereal grain in Northern Europe due to climatic change. *Food Addit Contam A* 2012;29:1543–1555.

[54] Kokkonen M, Ojala L, Parikka P, Jestoi M. Mycotoxin production of selected *Fusarium* species at different culture conditions. *Int J Food Microbiol* 2010;143:17–25.

[55] Elen O, Klemsdal SS, Clasen P-E, Razzaghian MJ. *Fusarium* spp. and mycotoxins in Norwegian wheat grain (2001–2004). In: *Abstracts, XV Congress of European Mycologists, Sept 16–21 2007, St. Petersburg, Russia*; 2007. p. 275–276.

[56] Imathiu SM, Hare MC, Ray RV, Back M, Edwards SG. Evaluation of pathogenicity and aggressiveness of *F. langsethiae* on oat and wheat seedlings relative to known seedling blight pathogens. *Eur J Plant Pathol* 2010;126:203–216.

[57] Waalwijk C, van der Lee T, de Vries I, Hesselink T, Arts J, Kema GHJ. Synteny in toxigenic *Fusarium* species: The fumonisin gene cluster and the mating type region as examples. *Eur J Plant Pathol* 2004;110:533–544.

[58] Nielsen LK, Jensen JD, Nielsen GC et al. *Fusarium* head blight of cereals in Denmark: Species complex and related mycotoxins. *Phytopathology* 2011;101:960–969.

[59] Pettersson H, Borjesson T, Persson L, Leirenius C, Berg G, Gustafsson G. T-2 and HT-2 toxins in oats grown in Northern Europe. *Cereal Res Commun* 2008;36B:591–592.

[60] Edwards SG. *Fusarium* mycotoxin content of UK organic and conventional oats. *Food Addit Contam A* 2009;26:1063–1069.

[61] Miedaner T, Cumagun CJR, Chakraborty S. Population genetics of three important head blight pathogens *Fusarium graminearum*, *F. pseudograminearum* and *F. culmorum*. *J Phytopathol* 2008;156:129–139.
[62] Stępień L, Chełkowski J. *Fusarium* head blight of wheat: pathogenic species and their mycotoxins. *World Mycotox J*. 2010;3:107–119.
[63] Strub C, Pocaznoi D, Lebrihi A, Fournier R, Mathieu F. Influence of barley malting operating parameters on T-2 and HT-2 toxinogenesis of *Fusarium langsethiae*, a worrying contaminant of malting barley in Europe. *Food Addit Contam A* 2010;27:1247–1252.
[64] Lukanowski A, Sadowski C. *Fusarium langsethiae* on kernels of winter wheat in Poland – Occurrence and mycotoxigenic abilities. *Cereal Res Commun* 2008;36B:453–457.
[65] Stanković S, Lević J, Ivanović D, Krnjaja V, Stanković G, Tančić S. Fumonisin B_1 and its co-occurrence with other fusariotoxins in naturally-contaminated wheat grain. *Food Control* 2012;23:384–348.
[66] Ivić D, Domijan AM, Peraica M, Miličević T, Cvjetković B. *Fusarium* spp. contamination of wheat, maize, soybean, and pea grain in Croatia. *Arh Hig Rada Toksikol* 2009;60:435–442.
[67] Infantino A, Pucci N, Conca G, Santori A. First report of *Fusarium langsethiae* on durum wheat kernels in Italy. *Plant Dis* 2007;91:1362.
[68] Ward TJ, Bielawski JP, Kistler HC, Sullivan E, O'Donnell K. Ancestral polymorphism and adaptive evolution in the trichothecene mycotoxin gene cluster of phytopathogenic *Fusarium*. *Proc Natl Acad Sci USA* 2002;99:9278–9283.
[69] Desjardins AE, Jarosz AM, Plattner RD, Alexander NJ, Brown DW, Jurgenson JE. Patterns of trichothecene production, genetic variability, and virulence to wheat of *Fusarium graminearum* from smallholder farms in Nepal. *J Agric Food Chem* 2004;52:6341–6346.
[70] Jennings P, Coates ME, Walsh K, Turner JA, Nicholson P. Determination of deoxynivalenol- and nivalenol-producing chemotypes of *Fusarium graminearum* isolated from wheat crops in England and Wales. *Plant Pathol* 2004;53:643–652.
[71] Qu B, Li HP, Zhang JB et al. Comparison of genetic diversity and pathogenicity of fusarium head blight pathogens from China and Europe by SSCP and seedling assays on wheat. *Plant Pathol* 2008;57:642–651.
[72] Schmale DG, Wood-Jones AK, Cowger C, Bergstrom GC, Arellano C. Trichothecene genotypes of *Gibberella zeae* from winter wheat fields in the eastern USA. *Plant Pathol* 2011;60:909–617.
[73] Zhang H, van der Lee T, Waalwijk C et al. Population analysis of the *Fusarium graminearum* species complex from wheat in China shows a shift to more aggressive isolates. *PLoS ONE* 2012;7(2):e31722. Doi:10.137/journal.pone.0031722.
[74] Karugia GW, Suga H, Gale LR, Nakajima T, Tomimura K, Hyakumachi M. Population structure of the *Fusarium graminearum* species complex from a single Japanese wheat field sampled in two consecutive years. *Plant Dis* 2009;93:170–174.
[75] Lee SH, Lee J, Nam YJ, Lee S, Ryu JG, Lee T. Population structure of *Fusarium graminearum* from maize and rice in 2009 in Korea. *Plant Pathol* 2010;26:321–327.
[76] Madgwick JW, West JS, White RP et al. Impacts of climate change on wheat anthesis and fusarium ear blight in the UK. *Eur J Plant Pathol* 2011;129:117–131.
[77] Zhang X, Halder J, White RP et al. Climate change increases risk of fusarium ear blight on wheat in central China. *Ann Appl Biol* 2014;164:384–395.
[78] Moschini RC, Pioli R, Carmona M, Sacchi O. Empirical predictions of wheat head blight in the northern Argentinian Pampas region. *Crop Sci* 2001;41:1541–1545.
[79] De Wolf ED, Madden LV, Lipps PE. Risk assessment models for wheat fusarium head blight epidemics based on within-season weather data. *Phytopathology* 2003;93:428–435.
[80] Rossi V, Giosuè S, Pattori E, Spanna F, Del Vecchio A. A model estimating the risk of Fusarium head blight on wheat. Bull *OEPP* 2003;33:421–425.

[81] Del Ponte EMD, Fernandes JMC, Pavan W. A risk infection simulation model for fusarium head blight of wheat. *Fitopatol Bras* 2005;30:634–642.

[82] Xu XM, Monger W, Ritieni A, Nicholson P. Effect of temperature and duration of wetness during initial infection periods on disease development, fungal biomass and mycotoxin concentrations on wheat inoculated with single, or combinations of, *Fusarium* species. *Plant Pathol* 2007;56:943–956.

[83] West JS, Holdgate S, Townsend JA, Edwards SG, Jennings P, Fitt BDL. Impacts of changing climate and agronomic factors on fusarium ear blight of wheat in the UK. *Fungal Ecol* 2012;5: 53–61.

[84] Xu XM, Nicholson P. Community ecology of fungal pathogens causing wheat head blight. *Ann Rev Phytopathol* 2009;47:83–103.

Leif Sundheim and Trond Rafoss

6 Fumonisin in maize in relation to climate change

6.1 Introduction

In 1904 the American plant pathologist Sheldon described *Fusarium moniliforme* J. Sheld. from moldy maize, which was suspected of causing animal diseases in the state of Nebraska, USA [1]. *Fusarium moniliforme* is today considered a synonym of *F. verticillioides* (Sacc.) Nirenberg [2].

Fig. 6.1: Maize infected with *Fusarium verticillioides*. Photo: Hadush Tsehaye.

Maize infected with *F. verticillioides* was implicated in several human and animal diseases in the twentieth century. Esophageal cancer of humans was associated with consumption of maize infected with *F. verticillioides* in a region of South Africa [3]. Outbreaks of a serious horse disease, equine leukoencephalomalacia (ELEM), in South Africa during the 1970s led to an intensified search for the causal agent, and feeding trials with infected maize confirmed the suspicion that maize contaminated with *F. verticillioides* resulted in ELEM in horses [4]. In 1988 South African scientists from the "Programme on Mycotoxins and Experimental Carcinogenesis" (PROMEC) isolated a toxin from *F. verticillioides* cultures which represented a new class of fungal toxins. Gelderblom et al. named the toxin fumonisin [5], and Bezuidenhout et al. published its chemical structure [6]. Marasas et al. demonstrated that fumonisin B_1 (FB_1) caused ELEM in horses in South Africa [7], and Rheeder et al. reported on higher levels of the fumonisins FB_1 and FB_2 in maize samples from areas with high rates of human esophageal cancer than in areas with low rates of that cancer in the country [8]. Swine in the USA fed maize infected with *F. verticillioides* died of porcine pulmonary edema and hydrothorax, and fumonisin B_1 (FB_1) was shown to be the causal agent [9]. The high incidence of human neural tube defects in Texas, USA, was associated with fumonisin-contaminated maize [10].

The fumonisins are grouped in A, B, C, and P series, and FB_1 is the most common and most toxic fumonisin. The members of the FB series have been most intensively studied because of their toxicity and potential carcinogenicity. The highest levels of FB_1 toxin occur in moldy maize and maize-based food and animal feed [11]. Several research groups have studied the phytotoxicity of FB_1. Naik et al. found no effect of *F. verticillioides* on maize seed germination, emergence, and maize yield [12]. Doehlert et al. reported that, while 100 ppm FB_1 had no effect on germination of maize seed, the toxin inhibited radicle elongation [13]. Lamprecht et al. found that fumonisins caused dose-dependent reductions in shoot and root length and dry mass of maize and tomato seedlings, and the authors concluded that FB_1 was more phytotoxic to seedlings than FB_2 and FB_3 [14].

The fumonisin FB_1 was classified as a group 2B carcinogen (probably carcinogenic in humans) in 1993 by the International Agency for Research on Cancer [15]. The association between consumption of fumonisin-contaminated maize and a high incidence of human esophageal cancer in South Africa was one of the reasons for the classification [8]. In Linxian County, a high-risk area for esophageal cancer in China, 80 maize samples from local households contained high levels of fumonisins [16]. In Iran, elevated fumonisin levels in maize have been reported in an area with high esophageal cancer frequency [17].

6.2 Fumonisin-producing fungi

The teleomorph of *F. verticillioides* was described as *Gibberella fujikuroi* (Sawada) Wollenw. in 1931 [18]. In currently accepted nomenclature, *F. verticillioides* is the anamorph mating population A of *G. fujikuroi* [19]. In maize, *F. verticillioides* and *F. proliferatum* (Matsush.) Nirenberg are the major fumonisin producers because both species are widely distributed and yield high levels of the toxin [20]. Under optimal conditions *in vitro* a selected strain of *F. verticillioides* produced 3 to 4 g of FB_1 per kg of culture material [21]. Following the elucidation of the structure of FB_1, the fumonisins FB_2 and FB_3 were discovered, and these three are the major fumonisins in maize naturally infected by *F. verticillioides* and *F. proliferatum* [20]. Only the FB series occurs at significant levels in naturally contaminated maize in the field, while fumonisins of the A, C, and P series are minor metabolites produced on synthetic media in the laboratory [22].

A total of 15 *Fusarium* spp. are known to produce fumonisins [20, 23]. In addition to *F. verticillioides* and *F. proliferatum* there are 13 known fumonisin producers within the genus *Fusarium* [20]. Less than 50 % of the isolates from each of these 13 species have been shown to produce fumonisins [23]. Only trace amounts of fumonisins have been reported from strains of *F. andiyazi* and *F. pseudonygamai* [20].

Fumonisin-producing species within the genus *Fusarium*:

- *F. andiyazi* Marasas, Rheeder, Lampr., K. A. Zeller, and J. F. Leslie
- *F. anthophilum* (A. Braun) Wollenw.
- *F. dlaminii* Marasas, P. E. Nelson, and Toussoun
- *F. fujikuroi* Nirenberg
- *F. globosum* Rheeder, Marasas, and P. E. Nelson
- *F. napiforme* Marasas, P. E. Nelson, and Rabie
- *F. nygamai* L. W. Burgess, and Trimboli
- *F. oxysporum* Schltdl.
- *F. polyphialidicum* Marasas, P. E. Nelson, Toussoun, and P. S. van Wyk
- *F. proliferatum* (Matsush.) Nirenberg
- *F. pseudonygamai* O'Donnell, and Nirenberg
- *F. sacchari* (E. J. Butler and Hafiz Khan) W. Gams
- *F. subglutinans* (Wollenw. and Reinking) P. E. Nelson, Toussoun, and Marasas
- *F. thapsinum* Klittich, J. F. Leslie, P. E. Nelson, and Marasas
- *F. verticillioides* (Sacc.) Nirenberg

In Denmark, Frisvad et al. discovered that the industrially important, saprophytic fungus *Aspergillus niger* Thiegh. produced low levels of FB_2 on synthetic media [24]. From a collection of 180 industrial and non-industrial *A. niger* strains, as much as 81 % of the strains produced FB_2, FB_4 or FB_6 [25]. One *A. niger* strain used for industrial citric acid production yielded 3.3 ppm fumonisin in broth culture, while among 17 other black *Aspergillus* species none produced fumonisin [25]. In Portugal, 270 *A. niger* strains were isolated from 73 maize samples, and 39 % of the strains produced FB_2 *in vitro*. The mean *in vitro* FB_2 production of *A. niger* strains obtained in three geographical regions in Portugal were from 0.37 to 0.52 $\mu g\,g^{-1}$ [26].

6.2.1 Biology of fungi producing fumonisin

Maize is the crop most often contaminated with fumonisin, but a limited number of other crop plants and commodities may also contain the toxins [27, 28]; see Fig. 6.1. The major fumonisin producers *F. verticillioides* and *F. proliferatum* are important maize pathogens causing seedling blight, root rot, stalk rot, and ear rot [28–30]. Munkvold et al. compared methods for inoculation of maize plants with *F. verticillioides* strains labeled as nitrate-nonutilizing (*nit*) mutants [31]. Seed inoculation resulted in systemic infection of plants and kernels, but local inoculation of the silks resulted in more kernel infection than systemic growth of the fungus from infected kernels [31].

Survival of *F. verticillioides* in the field has not been studied to the same extent as for other *Fusarium* spp. In most countries maize is grown in fields where the crop was cultivated the previous year or in short rotations with other crops. Studies by Cotton and Munkvold demonstrated that *F. verticillioides* survives in maize stalks in the field for at least 630 days, and survival after one year in maize stalks buried 30 cm deep was equivalent to survival on the soil surface [32]. *Fusarium verticillioides* and *F. proliferatum* rarely develop perithecia, but they produce large amounts of microconidia and macroconidia, which are the most important inoculum for the fungi [28, 30].

Systemic infection by *F. verticillioides* in maize was first studied by Foley [33]. He isolated the fungus from symptom-free stalks, leaf sheaths and axillary buds. Isolation frequency was highest from leaf sheaths, and successful isolations of *F. verticillioides* from nodes were more frequent than isolation from internodes of maize [33]. The endophytic growth of *F. verticillioides* was further investigated by Munkvold et al. [34]. They inoculated seeds and followed transmission of the fungus to stalks and developing kernels. In three field experiments seed infection had no significant effect on the percentage of crown, stalk, and kernel infected by *F. verticillioides* [34]. Silk inoculation with the fungus was the only method which significantly increased kernel infection, and on individual ears up to 100 % of the kernels became infected. The mean result from silk inoculation was 83.7 % infected kernels [34]. In a controlled environment *F. verticillioides* developed more rapidly at high temperature than at moderate temperature during the vegetative stage, but temperature did not affect the incidence of kernel or ear infection [35]. Three *F. verticillioides* strains, which were transformed with the *gusA* reporter gene, maintained their ability to infect maize, and the transformed strains facilitated study of the endophyte-maize relationship [36]. Mycelia with GUS (β-glucuronidase) activity were recovered from roots on maize seedlings developing from kernels infected with isolates carrying the *gusA* gene [36].

Both the fumonisin-producing *F. verticillioides* and the DON-producing *F. graminearum* are common maize pathogens, and the two fungi frequently compete in maize plants. Reid et al. inoculated maize ears with mixtures of *F. verticillioides* and *F. graminearum*, and DNA analysis revealed that *F. verticillioides* had the greater growth rate of the two pathogens [37]. The FB_1 level was not different when the two fungi were applied together to when *F. verticillioides* was applied alone [37]. Based on single, mixed, and sequential inoculations with the two fungi Picot et al. concluded that previous contamination by *F. graminearum* can facilitate subsequent infection by *F. verticillioides* [38]. In southern Europe *F. verticillioides* and *F. proliferatum* frequently occur in maize [39].

Both environmental factors and maize genotype determine infection by fumonisin-producing *Fusarium* spp. and accumulation of the mycotoxin in maize. Field trials with 15 maize genotypes at 17 locations in the USA revealed a highly significant hybrid × location interaction [40]. The mean fumonisin content in the maize hybrids varied from 0.5 to 48.5 $\mu g\ g^{-1}$, and there was an increase in fumonisin content as the latitude of the location decreased [40]. Desjardins and Plattner inoculated maize with

F. verticillioides strains with different fumonisin production potential to study interaction in field experiments [41]. There was no consistent difference in fumonisin content of maize following inoculation with low fumonisin-producing strains and inoculation with strains with high fumonisin production [41]. Doko et al. determined fumonisin levels in 26 maize genotypes grown in three European and two African countries, and found a positive correlation between high FAO maturity class of the genotype and fumonisin contamination [42]. To develop methods for *F. verticillioides* inoculation of maize in a breeding program, Clements et al. compared injection of inoculum through the ear husk leaves, spray inoculum twice on the silk with ear subsequently covered by a bag, and insertion of toothpicks infected with *F. verticillioides* [43]. Only injection through the husk leaves significantly increased severity of *Fusarium* ear rot and the concentration of fumonisin in grain compared to the control [43].

6.3 Fumonisin accumulation in developing maize kernels

In field trials with three maize hybrids at two locations in North Carolina, USA, Bush et al. sampled maize kernels weekly to follow *F. verticillioides* infection and FB_1 accumulation [44]. FB_1 contamination appeared as kernels neared physiological maturity four weeks after pollination and increased up to the harvest date. Peak FB_1 content varied from 5 to 10 $\mu g\, g^{-1}$ in the hybrids [44]. The authors suggested that early maize harvest at > 25 % grain moisture may reduce the level of fumonisin contamination in years with heavy infection [44].

In Japan Uegaki et al. studied the increase in fumonisin concentration during growth of the maize plant [45]. Maize planted in the field in mid-May was sampled every two weeks from mid-June. The fumonisin concentration in maize kernels increased abruptly after the dough-ripe stage and kept increasing until the full-ripe stage [45]. The mean concentration of FB_1 in the ears exceeded 80 % of final fumonisin content in mid-September, and continued to increase to the full ripe stage in late October, when the mean concentration was 2.3 $\mu g\, g^{-1}$ [45].

In field trials at two locations in France maize was inoculated with *F. verticillioides* four days after silking [46]. Samples for *F. verticillioides* DNA and fumonisin (FB_1, FB_2, and FB_3) analysis were collected at intervals. Fungal DNA was detected seven days after inoculation, while fumonisin was first detected 15 days after inoculation and continued to increase up to 42 days after inoculation. The authors concluded that the dent stage is the most significant period for fumonisin biosynthesis in maize kernels [46].

6.3.1 Fumonisins are not required for pathogenicity

The *F. verticillioides* genome has been partly sequenced. The fumonisin biosynthesis cluster of the genome consists of several *FUM* genes which are required for fumonisin biosynthesis [47]. There are 17 FUM genes known, and evidence for epigenetic regulation of *FUM* genes in *F. verticillioides* was presented by Visentin et al. [48]. Fumonisin accumulation correlates with the amount of transcript from the key genes, *FUM1*, *FUM21*, and *FUM8* [48].

Proctor et al. developed fumonisin nonproducing mutants which were virulent on maize [49]. Transformation of the *FUM5* gene into a fumonisin nonproducing mutant restored fumonisin production, and disruption of the *FUM5* gene reduced fumonisin production by 99 % [50]. *Fusarium verticillioides* strains which produced only one of the fumonisins (FB_1, FB_2, or FB_3) and a fumonisin nonproducing strain did not lose their ability to cause ear rot of maize [51]. Desjardins and Plattner concluded that these results indicate that loss of fumonisin production does not affect the ability of *F. verticillioides* to cause maize ear rot [51].

6.3.2 Insect damage increases risk of fumonisin contamination

Several insect species are serious pests on maize ears. In USA, the European corn borer (ECB; *Ostrinia nubilalis*) larvae develop in the stalks and ears of maize. The first generation infests plants during the vegetative stage, while adults of the second generation lay egg masses on leaf surfaces, and the larvae feed on leaves, stalks, and ears [52].

Maize hybrids were genetically modified to express *Bacillus thuringiensis* genes coding for the CryIAb protein, which give a high level of resistance to feeding by ECB larvae. In field trials conducted over three years Munkvold et al. found that transgene maize hybrids expressing the CryIAb protein in kernels had less *Fusarium* ear rot compared to near-isogenic hybrids lacking the *cryIAb* gene [53]. In an area of France where maize was regularly infested with ECB and a maize borer (*Sesamia nonagrioides*), Folcher et al. found 90 % reduction in concentration of fumonisin in maize genetically modified for resistance to ECB compared to nonmodified isogenic lines [54].

The CryIAb protein is not effective against the Western bean cutworm (WBC; *Striacosta albicosta*), corn earthworm (CEW; *Helicoverpa zea*) and some other Lepidopteran maize pests [55]. In field trails in Iowa, USA, Bowers et al. planted maize hybrids genetically modified to express both the ViP3Aa protein and the CryIAb protein for protection against several Lepidopteran pests. Natural infestation by WBC, ECB, and CEW was the treatment in the experiments [55]. The insect damage was reduced and the mean fumonisin content was 0.56 $\mu g\, g^{-1}$ in maize hybrids expressing the two proteins CryIAb and ViP3Aa, while the mean fumonisin contamination of nontransgenic maize hybrids was 5.47 $\mu g\, g^{-1}$ [55].

Insect infestation may be stimulated by *F. verticillioides* infection. In inoculation experiments at the International Institute of Tropical Agriculture (IITA), Nigeria, Cardwell et al. found that *F. verticillioides*-inoculated maize had higher levels of Coleopteran beetles and Lepidopteran borers than non-inoculated maize [56].

6.3.3 Small grain cereals contaminated with fumonisins

Fusarium spp. cause ear rot, stalk rot, and yield losses in sorghum (*Sorghum bicolor*) and pearl millet (*Pennisetum glaucum*), but fumonisin contamination of these African staple food crops has been less studied than in maize. Leslie et al. determined fumonisin production in five strains of each of five *Fusarium* spp. isolated from sorghum and pearl millet grown in Africa and Australia [57]. The two species, *F. nygamai* and *F. verticillioides*, produced high levels of FB_1 and FB_2, while strains of *F. andiyazi*, *F. pesudonygamai*, and *F. thapsinum* yielded only low levels or no fumonisins [57].

In a two-year survey of durum wheat in Italy, Palacios et al. found *F. proliferatum* to be the most common *Fusarium* species, with infection frequencies ranging from 8 % to 66 % [58]. During the 2007 season the fumonisin contamination (FB_1+ FB_2) levels varied from 0.01 to 1.25 $\mu g\,g^{-1}$, while only very low levels were detected in samples from 2008 [58]. Desjardins et al. reported that in wheat grown in Nepal, kernel black point disease and fumonisins were associated with *F. proliferatum* strains [59]. In inoculation experiments *F. proliferatum* reduced yield, and one strain obtained from wheat grown in Nepal produced 49 $\mu g\,g^{-1}$ of fumonisins (FB_1, FB_2, and FB_3) on wheat [59]. When Busman et al. analyzed wheat samples with wheat kernel black point disease from across the USA, most samples did not contain fumonisins above the level of detection (LOD), while low levels of fumonisins were detected in 9 of 43 samples [60]. Five samples contained > 0.05 $\mu g\,g^{-1}$ (FB_1, FB_2, and FB_3), including one sample with 2.2 $\mu g\,g^{-1}$ and one sample with 1.2 $\mu g\,g^{-1}$ (FB_1, FB_2, and FB_3) [60]. *Fusarium proliferatum* was the dominant *Fusarium* species,but the low levels of fumonisin in the samples provided evidence that fumonisins are not likely to contribute to the ability of *Fusarium* spp. to cause kernel black point disease of wheat in USA [60].

Thailand is the leading rice export country in the world. In a survey, Tansakul et al. analyzed unpolished samples of Thai red cargo rice from the retail market, and only 2 of 58 samples were found to be contaminated with fumonisins above LOD, both with only trace levels (< 5.0 ng g^{-1}) of FB_1 , and no FB_2 [61]. Two years' analysis of rice samples sold in Canada which originated from several countries including the USA, Canada, Pakistan, India, and Thailand revealed fumonisins in 15 of 99 samples the first year, and in 1 of 100 samples the second year [62]. The average concentration of FB_1 was 4.5 ng g^{-1} in the positive samples. In addition, FB_2 and FB_3 were detected in concentrations of > 1 ng g^{-1} [62].

6.3.4 Other crops and commodities contaminated with fumonisins

Noonmin et al. detected FB_2 in coffee bean samples of *Coffea arabica* and *C. canephora* grown in Thailand [63]. No *Fusarium* species known to produce fumonisin were identified in the coffee, but strains of *Aspergillus niger* isolated from the coffee beans produced low levels of FB_2 and FB_4 in a culture medium [63]. Very low levels of FB_2 (1.0–

9.7 ng g^{-1}) were detected in seven of twelve coffee bean samples analyzed [63]. All of the 50 randomly selected strains of *F. proliferatum* isolated from asparagus crowns in China produced FB_1 and FB_2 in maize grain cultures and on asparagus spear extracts, and FB_3 was isolated from most of the culture extracts as well [64]. Marín et al. did inoculation experiments on edible pine nuts (*Pinus pinea*) with eleven strains of *F. proliferatum* isolated from pine nuts, and identified six strains which produced fumonisins in shelled pine nuts and two strains which produced fumonisins in whole pine nuts [65]. The FB_1 levels in inoculated pine nuts varied from 0.10 to 0.69 µg g^{-1} [65].

Stratakou et al. conducted a survey to determine the level of mycotoxins in European grapes and wines, reporting on frequent occurrence of low levels of fumonisins in European wine [66]. Logrieco et al. analyzed 51 market samples of Italian red, white, and rosé wines, and reported that 9 of the 45 red wine samples contained FB_2 at levels ranging from 0.4 to 2.4 ng ml^{-1}, while FB_4 was not detected in any of the samples [67]. Mogensen et al. analyzed 77 wine samples from 13 countries, and 23 % of the samples contained FB_2 in the range of 1 to 25 ng ml^{-1} [68]. Powdery mildew caused by the fungus *Erysiphe necator* is one of the most common grape diseases. Damage to the berry surface by powdery mildew infection was identified as an infection site for strains of *A. niger* producing FB_2 in grapes [69]. Black *Aspergillus* strains were isolated from grapes in Argentina, and fumonisin production was assessed on a Czapek yeast medium. The only strains able to produce fumonisins were five *A. niger* strains, while strains of *A. carbonarius*, *A. japonicas*, and *A. tubingensis* did not produce fumonisins on the medium [70].

The fumonisins are water soluble and withstand the beer production process, which includes malting, mashing, boiling, and fermentation. In a survey of 29 domestic and imported beer brands marketed in the United States, 86 % contained FB_1 and 41 % contained FB_2 [71]. The total fumonisins (FB_1 and FB_1) varied from 0.3 to 12.7 ng ml^{-1}, with a mean concentration of 4.0 ng ml^{-1} in positive samples [71]. Samples of home brewed beer in South-Africa all contained fumonisins (FB_1, FB_2 and FB_3), and the mean concentration was 0.37 µg ml^{-1} (range 0.04–1.33 µg ml^{-1}; [72]). Transfer of FB_1 from naturally contaminated malted barley and maize grit to beer was studied by Pietri et al. in Italy (73]. Beer production based on maize grit contaminated with 1.1 to 3.2 µg g^{-1} FB_1 was followed, and the measured concentrations of FB_1 in the brewed beer varied from 37 to 89 ng ml^{-1}, while FB_2 was not found in the beer samples [73].

6.4 Geographical distribution of fumonisins in maize

Maize is widely grown in tropical, subtropical, and temperate climates, and cultivation of the crop is expanding. The yield potential has increased, and the current maize harvest weight is larger than any other cereal crop. In maize growing countries the crop is commonly contaminated with fumonisin [22].

6.4.1 Africa

Fandohan et al. reviewed the fumonisin situation for maize in Africa [74]. While maize is an important crop in Africa, only limited information on fumonisin contamination is available, except for South Africa, and there is a need for increased monitoring of maize in the continent [74].

The mycotoxin situation of food crops in Sub-Saharan Africa was summarized by Bankole et al., who concluded that fumonisins are the most important *Fusarium* mycotoxins in Sub-Saharan Africa [75]. Maize fumonisin contamination is widespread in West Africa, and contamination levels are frequently high. Maize samples from Nigeria had an FB_1 incidence of 78.6 % and a concentration range of 0.07 to 1.78 $\mu g\, g^{-1}$, with a mean concentration of 0.50 $\mu g\, g^{-1}$, while FB_2 was detected in 66 % of the samples with a mean concentration of 0.11 $\mu g\, g^{-1}$ [76]. All 32 maize samples collected on maize farms in Burkina Faso contained fumonisin in the range of 0.11 to 3.12 $\mu g\, g^{-1}$, and the mean fumonisin content was 1.17 $\mu g\, g^{-1}$ [77]. Analysis of 72 maize samples from a large Burkina Faso market revealed a 100 % fumonisin incidence and a mean contamination of 2.90 $\mu g\, g^{-1}$ with a range of 0.13 to 16.04 $\mu g\, g^{-1}$ [77]. *Fusarium verticillioides*, with an incidence of 68 %, and *F. proliferatum*, with an incidence of 31 %, were the most common *Fusarium* spp. isolated from maize in Benin, and the FB_1 and FB_2 levels varied from LOD to 6.7 $\mu g\, g^{-1}$ [78]. Maize was analyzed at harvest and during storage in different regions of Benin. The fumonisin levels were lowest in the Sudan Savannah region [78]. Rice, maize, and groundnuts were analyzed in Côte d'Ivoire for mycotoxins, but only low levels of fumonisin were detected [79].

Ethiopian scientists analyzed barley, wheat, sorghum, and teff (*Eragrostis tef*) for mycotoxins, but only detected fumonisins in sorghum samples, and the maximum concentration was 2.12 $\mu g\, g^{-1}$ [80]. Maize is the major cereal crop in Uganda and an important export commodity. *Fusarium* and fumonisins in maize grain stored in traditional structures were analyzed by Atukwase et al., who found that all maize samples analyzed contained fumonisins in the range of 0.27 to 10.0 $\mu g\, g^{-1}$ [81]. The *Fusarium* incidence level increased after harvest from 61.9 % to 77.5 % during the first two months of storage and then decreased to 31.9 % after sixth months of storage in Uganda [82]. Fumonisin levels decreased at one site in Uganda from an average of 5.7 $\mu g\, g^{-1}$ at harvest to 2.8 $\mu g\, g^{-1}$ after six months of storage [82].

In Kenya *F. verticillioides* is the dominant cause of maize ear rot [83]. 70 isolates of *F. verticillioides* were obtained from maize samples collected in the important maize growing regions of the country, and *in vitro* 74 % of the isolates produced fumonisin B_1 in the range of 0.07 to > 5.0 $\mu g\, g^{-1}$, with a mean of 1.51 $\mu g\, g^{-1}$ [83]. Analysis of 20 inbred maize lines from Zambia gave a median concentration of 0.10 $\mu g\, g^{-1}$ (FB_1 and FB_2) with a variation from 0.02 to 1.71 $\mu g\, g^{-1}$ (FB_1 and FB_2) [42]. In field experiments in Zambia with maize hybrids artificially inoculated with *F. verticillioides*, the mean concentration content of (FB_1 and FB_2) was 0.67 $\mu g\, g^{-1}$ in the harvested maize, with a variation from LOD to 13.05 $\mu g\, g^{-1}$ (FB_1 and FB_2) [84]. A survey in Zambia suggested that subsistence farmers and consumers might be exposed to dangerous levels of fumonisins, as in six districts the concentration was ten times higher than the level recommended by FAO/WHO [85]. Maize samples in Egypt were analyzed for fumonisin over a period of two years. The mean FB_1 content for the 2012 samples was 0.5 µg g^{-1} and 0.98 µg g^{-1} for the 2013 samples. The range of FB_1 content in the samples was 0.17 to 1.9 µg g^{-1} [86].

6.4.2 Europe

The maximum level of fumonisins in the EU is regulated by the Commission Regulation (EC) No 1126/2007 at 2.0 µg g^{-1} (FB_1 and FB_2) for unprocessed maize, 1.0 µg g^{-1} (FB_1 and FB_2) for maize flour, maize meal, maize grits, maize germ, and refined maize oil, 0.4 µg g^{-1} (FB_1 and FB_2) in maize-based food intended for direct human consumption, and 0.2 µg g^{-1} (FB_1 and FB_2) in processed maize-based foods and baby foods for infants and young children [87]. Visconti and Doko assayed fumonisin production in 28 strains of *F. verticillioides* and one strain of *F. proliferatum* from Italy, Spain, and France, and found that all strains produced FB_1 [88]. Logrieco et al. reported that the fumonisin-producing species *F. verticillioides* and *F. proliferatum* are commonly isolated from maize ear rot in southern Europe [39]. Up to 80 % of ears damaged by the corn borer (*Pyralis nubilalis*) are infected [39]. Ariño et al. did not find any difference in fumonisin content between tillage systems, type of irrigation, and harvest date in Spain, while higher levels of nitrogen fertilizers tended to increase fumonisin content in maize and cultivation of insect resistant varieties reduced fumonisin levels [89]. Binder et al. found fumonisin in 19 of 26 maize samples analyzed from Europe, with a median fumonisin concentration of 0.43 µg g^{-1} [29]. Ivic et al. isolated *F. verticillioides* from barley grain in Bosnia and Herzegovina, but the average FB_1 content in barley grain was very low, with an average of 2.40 ng g^{-1}, with a range of 1.01 to 5.35 ng g^{-1} in the samples analyzed [90].

The incidence of *F. verticillioides* and *F. proliferatum* is commonly low in the maize growing countries of central and northern Europe [91]. Van Asselt et al. presented data from mycotoxin monitoring in silage maize from the Netherlands, Belgium, Germany, and France during the period 2003–2007 [91]. High fumonisin concentrations were found in 2003 and 2007, indicating of suitable conditions for infection by *F. verticillioides* and *F. proliferatum* in Western Europe in two of five recent years. The mean fumonisin contents of positive samples were 1.55 µg g^{-1} in 2003 and 0.84 µg g^{-1} in 2007, demonstrating that west European maize crops may contain fumonisin above the EU maximum level of 2.0 µg g^{-1} [87, 91]. No fumonisin was detected in 12 samples in 2005 and only very low levels in 2004 and 2006 [91]. In France maize products in horse feed led to two cases of ELEM in the region of Toulouse, and Bailly et al. analyzed the feed and found that it contained 125 µg g^{-1} FB_1 [21].

Maize is currently cultivated for silage in Denmark and in the Baltic states. Even in the best climatic zones of Norway, Sweden, and Finland growers have successfully grown maize during recent years. The species *F. verticillioides* and *F. proliferatum* were not detected in a recent Danish survey [92]. Higher temperatures will stimulate more maize cultivation, and warmer climate may increase the risk for fumonisin contamination in maize by *F. verticillioides* in the Nordic and Baltic countries [93].

6.4.3 South America

An outbreak of ELEM in the Buenos Aires province, Argentina, in 2007 was caused by contaminated maize. The dead horses had the symptoms of ELEM, and analyses revealed a content of 12.49 $\mu g\, g^{-1}$ FB_1 and 5.25 $\mu g\, g^{-1}$ FB_2 in the horse feed [94]. Fumonisins were determined in maize samples from the Entre Rios Province, Argentina, for two consecutive years, and in 2003 a total of 14 samples were analyzed. The median fumonisin content of the 12 positive samples was 5.80 $\mu g\, g^{-1}$, while seventeen samples were analyzed in 2004 and the median content of the 12 positive samples was 2.40 $\mu g\, g^{-1}$[95]. The maximum level detected was 34.70 $\mu g\, g^{-1}$ [95]. Ono et al. detected fumonisins in 98 % of 150 samples from maize 62 hybrids in Brazil, and 62 % of the samples had < 5.0 $\mu g\, g^{-1}$ [96]. Fumonisin contamination was higher in maize samples originating in the northern and central western regions than in those from the central southern region of Brazil [96]. Stumpf et al. analyzed 29 maize kernel samples form the Rio Grande do Sul state of Brazil. FB_1 was detected in 58.6 % and FB_2 was detected in 37.9 % of the samples. The mean FB_1 level was 0.66 $\mu g\, g^{-1}$, and the mean FB_2 level was 0.42 $\mu g\, g^{-1}$ [97].

6.4.4 North America

Shelby et al. concluded that fumonisin risk is highest in the southern USA states [40]. A significant negative correlation between latitude and fumonisin contamination was found in trials with the same genotype at 17 locations in the country. The mean total fumonisin contamination of the hybrids ranged from 5.8 to 48.5 $\mu g\, g^{-1}$. Maize from the southern and central states up to Iowa, Illinois, and Kansas was more contaminated than maize from the upper mid-western states [40]. Hooker and Schaafsma analyzed fumonisin content of maize from Ontario, Canada, over eight years. The FB_1 concentration at maturity was highly variable between fields and from year to year. Between 17 % and 56 % of the maize samples contained > 1.0 $\mu g\, g^{-1}$ FB_1 over the years. Based on the analytical data the authors concluded that maize hybrid was more important for fumonisin contamination than climatic variation between years in Canada [98].

Fusarium species of the *Gibberella fujikuroi* complex from pearl millet and maize grown in Georgia, USA, were studied by Jurjevic et al. [99]. Of the crosses between pearl millet isolates of *Fusarium* spp. and *F. verticillioides*, 50.4 % were successful, and when crossed with *F. proliferatum* 10.1 % of the crosses succeeded [99]. The percentages of maize isolates compatible with *F. verticillioides* ranged from 70.2 % to 89.5 %. When maize samples collected in the survey over three years were analyzed, between 63 % and 91 % of the samples contained fumonisins, and the fumonisin levels ranged from 0.6 to 33.3 μg g^{-1}, while FB_1 and FB_2 were not detected in any of the 81 pearl millet samples [99]. Wu et al. maintained that the risk for fumonisin contamination is greater in warm and relatively dry areas of lower latitudes and altitudes than in cooler and more humid climates in high latitudes and altitudes [100].

6.4.5 Asia

Binder et al. analyzed maize samples from 10 countries in Asia and two countries of Oceania, and fumonisins were detected in 69 % of 309 maize samples [29]. The average fumonisin concentration in maize was 1.34 μg g^{-1}, and the median fumonisin concentration was 0.53 μg g^{-1} [29]. The highest fumonisin concentration was determined in a maize sample from China with 14.71 μg g^{-1} fumonisin [29]. Only 4 of 98 wheat samples were contaminated with fumonisin, with a median concentration of 0.24 μg g^{-1}, and of 122 soybean meal samples only 9 contained fumonisin with a median concentration of 0.26 μg g^{-1} [29]. Wei et al. analyzed FB_1 and FB_2 in maize samples originating in four provinces of China [101]. In 307 maize samples the incidence of fumonisins (FB_1 and FB_2) varied from 31.5 % in the Gansu province to 81.1 % in the Shandong province. The average (FB_1 and FB_2) concentration was 0.70 μg g^{-1} and the range was from ≤0.01 to 13.11 μg g^{-1}. Seventeen of the 307 samples had higher fumonisin content than the current EU-regulated maximum level for human consumption [101].

Very high fumonisin levels in maize were reported from areas with a high risk of esophageal cancer in the Golestan Province, Iran [16]. The mean FB_1 concentration in 66 maize samples was 223.64 μg g^{1}, and the mean FB_1 concentration in 66 rice samples was 21.59 μg g^{1}.There was a significant, positive correlation between risk of esophageal cancer and fumonisin content and rice grown in this region, but the authors did not find a significant correlation between fumonisin content of maize and risk of esophageal cancer [17]. Azizi and Rouhi analyzed biscuits and cookies obtained from supermarkets in Babol City, Iran for fumonisins [102]. All of the 30 biscuit and cookie samples contained fumonisins above the LOD. 28 contained < 2 μg g^{-1} total fumonisin, and 2 samples contained > 2 μg g^{-1} fumonisin with 2.3 μg g^{-1} total fumonisin the highest level recorded [102]. In a Japanese study of maize the mean concentration of FB_1 was 2.31 μg g^{-1} at the full ripe stage. [45]. In Vietnam 8 of 25 samples of maize for human consumption were contaminated with FB_1, and the level ranged from 0.4 to 3.3 μg g^{-1} [103].

6.5 Climate change predicted by IPCC

The Intergovernmental Panel on Climate Change (IPCC) concluded in its 2013 report that the troposphere has warmed since the mid-20th century [104]. There is medium confidence in the rate of global warming in the Northern Hemisphere and low confidence elsewhere. IPPC has medium confidence that the projected increase in global mean surface temperature during the next twenty years will be in the range of 0.3–0.7 °C [104]. According to IPPC there will be an increase in global mean surface temperature for the period 2081–2100 relative to 1986–2005 [104]. IPCC has very high confidence that the Arctic region will warm more rapidly than the global mean, and the mean warming over land will be larger than over ocean. It is virtually certain that there will be more frequent hot and fewer cold temperature extremes over most land areas on daily and seasonal timescales as the global mean temperature increases [104]. IPPC concluded that there is high confidence that global warming has increased crop production in some high latitude regions in China and Europe [104]. For maize and other major crops temperature increases of 2 °C or more above late 20th century levels without adaptation will negatively impact maize production in tropical and temperate regions [104].

According to the IPPC precipitation has increased since 1901 in the Northern Hemisphere, and high latitudes are likely to experience an increase in precipitation by the end of this century [104]. Confidence in precipitation changes averaged over global land areas since 1901 is medium prior to 1951 and high afterwards. Area-averaged long-term positive or negative precipitation trends have low confidence for other latitudes [104]. The frequency or intensity of heavy precipitation has increased in North America and Europe. Confidence in changes in precipitation is at most medium in other continents [104].

IPPC concluded that the level of CO_2 in the atmosphere has increased in the last three centuries. From an estimated concentration of 280 ppm CO_2 in the atmosphere around 1750, the concentration had increased to 391 ppm CO_2 by 2011 [104]. According to the IPPC, man-made emissions are projected to further increase CO_2 in the atmosphere [104].

6.5.1 Climate effects on fungi producing fumonisin in maize

While maize production has doubled over the last 40 years, the percentage of loss due to plant pests in the crop has not changed much, amounting to almost one third of attainable yield during the period 2001–2003 [105]. Coakley et al. concluded that the effects of climate change may differ depending on the global location [106]. The rates of pathogen development may change and the physiology of host-plant interaction may be altered. The effects of climate change on plant disease management may be less important than changes in land-use patterns, transgenic technologies, and availability of chemical pesticides [106].

De la Campa et al. used fumonisin analysis data of maize grown in Argentina and the Philippines to develop a model for the effects of climate, insect damage, and hybrid maize on fumonisin contamination at harvest [107]. The fumonisin content varied from 0.3 to 12.0μg g^{-1} in the data from Argentina and from 0.3 to 1.8 μg g^{-1} in the data from the Philippines. Weather or location explained 47 %, insect damage severity in mature ears explained 17 %, and hybrid explained 14 % of the variability in fumonisin contamination [107]. Schaafsma and Hooker analyzed maize grown in the province of Ontario, Canada, and concluded that weather accounted for 19 % of the variation in fumonisin contamination [108]. The levels of fumonisin in Canadian maize were rather low, as only 17 % to 56 % of the maize samples contained > 1.0 μg g^{-1} fumonisin. Schaafsma and Hooker emphasized that the *Fusarium*-crop interaction is more complex in maize than in wheat because of the variations in the maize flowering period, the important role of insect damage for the *Fusarium* infection in maize, and the rapid turnover in maize genotypes [108].

6.5.2 Effects of temperature

Marin et al. determined the optimum *in vitro* temperature for growth of *F. verticillioides* and *F. proliferatum*, and when growth at 25 °C and 30 °C was compared, *F. verticillioides* grew faster at 30 °C than at 25 °C, while 25 °C was the optimum temperature for growth of *F. proliferatum* [109]. Growth of the two fungi was determined over a range of temperatures from 5 °C to 40 °C on sterile layers of maize, and the optimum temperature for growth was 30 °C for both *F. verticillioides* and *F. proliferatum* [110]. Based on evaluation in 36 environments in Spain, Cao et al. found that high temperature during flowering increased fumonisin content of maize kernels, while ear damage by corn borer and hard rainfall favored fumonisin accumulation in the kernel drying period [111].

6.5.3 Effects of drought

The IPCC concluded that precipitation will increase at higher latitudes, but in the major maize producing areas of the world precipitation will not compensate for the temperature-driven increase in evaporation, which will result in more droughts for most maize producing countries [104].

Water activity (a_w) is a measure of available water in the substrate. Free water has a water activity of 1.0 a_w, which also characterizes 100 % relative humidity. Marin et al. concluded that *in vitro* water availability of 0.994–0.98 a_w is optimal for fumonisin production by *F. verticillioides* [109]. At temperatures that were not optimal for fungal growth fumonisin production is higher at lower a_w values. Maize starch production in the kernels decreases water availability in the late phase of maturation [112]. Jurado et al. reported that a_w of 0.937 reduced growth of *F. verticillioides*, but increased expression of the *FUM1* gene [112]. Thus, water stress may be an important factor for fumonisin accumulation in the final phase of maize kernel development [112]. Marín et al. concluded that water stress resulted in an induction of the *FUM1* gene in *F. verticillioides*, but *F. proliferatum* showed a stable expression of the *FUM1* gene regardless of water potential [113].

Based on two years of field trials in four different states of the USA, Parsons and Munkvold concluded that dry weather after pollination and thrips infestation of the ears increased the risk for fumonisin contamination [114]. Wu et al. concluded that in North-America fumonisin risk is higher in Texas and the south-eastern states than in central USA, but the fumonisins are the most common mycotoxins in all the corn belt states of the USA [100]. In Central and South America and Southeast Asia the risk for fumonisin contamination of maize is higher in lower elevation maize production areas than in maize produced at higher elevations [100]. There has been an increase in maize cultivation in southern and central Europe, and a northward expansion of maize cultivation in Europe has been predicted with the projected climate changes [115].

Shelby et al. did field experiments in the 11 major maize producing states of the USA, and mean fumonisin content for 15 hybrids varied from 0.5 to 48.5 $\mu g\, g^{-1}$ at the different locations [40]. June precipitation was negatively correlated with fumonisin content, and June rainfall was the most important factor for the fumonisin levels of maize hybrids in the USA [40]. Drought just prior to pollination creates physiological stress for the maize plant, which is likely to create good conditions for infection and fumonisin production by *Fusarium* spp. [40].

The effect of water deficit on accumulation of FB_1 and FB_2 in maize was studied in the state of São Paulo, Brazil [116]. 35 cultivars were planted at three locations and sampled for fumonisin analysis during two seasons. All samples were contaminated with fumonisins with a content of FB_1 from 0.10 to 43.80 $\mu g\, g^{-1}$, and the FB_2 content varied from 0.04 to 11.65 $\mu g\, g^{-1}$. Fumonisin concentration was negatively correlated with precipitation in the period from silking to kernel milky stage [116]. Precipitation in the month before harvest increased the fumonisin content of the maize crop. In the period from maturity to harvest there were positive correlations between fumonisin concentration and precipitation and between water surplus in the soil and fumonisin concentration [116]. 92 % of samples analyzed in the northern region of Brazil had a fumonisin content > 1.0 $\mu g\, g^{-1}$, while only 18.5 % of the samples from the central southern region were contaminated with a fumonisin content > 1.0 $\mu g\, g^{-1}$ [117]. Precipitation in the month preceding harvest in the southern region was 92.8 mm, whereas the northern region had 202 mm precipitation in the month before harvest [117].

Schjøth et al. did field experiments with maize in Zambia in the high rainfall zone (HRZ), which has an annual precipitation of 1000–1500 mm, and in the medium rainfall zone (MRZ) with an annual precipitation of 800–1000 mm [84]. Three commercial maize hybrids were inoculated by a mixture of six isolates of *F. verticillioides* obtained from Zambian maize at both sites. Combined FB_{1-2} concentration varied from LOD to 13.05 µg g^{-1}, with an overall mean concentration of 0.7 µg g^{-1}. Maize from the HRZ had a low incidence of FB_{1-2} positive samples (mean 41 %) which all, apart from one, contained FB_{1-2} < 0.5 µg g^{-1}. There was a higher incidence of FB_{1-2} positive samples (mean 97 %) in the MRZ, and higher concentrations (40 % of the samples > 1.0 µg g^{-1}) were recorded [84].

Studies by Herrera et al. in Spain showed that water activity of the grain at above 0.93 a_w (20.0 % grain moisture) was the critical condition for fumonisin production in mature grain [118]. The frequency of fumonisin detection in field samples was 48.3 %, and the positive samples contained from LOD to 5.82 µg g^{-1} fumonisin. High temperature during flowering combined with wet weather in the last month preceding harvest were the most important factors for fumonisin contamination of maize [118].

6.5.4 Effects of elevated CO_2 level

Fungi can tolerate increased CO_2 and the predicted levels of 800–1000 ppm CO_2 may not limit growth of fungi producing mycotoxins in crops [119]. Plant anatomy may be altered with increased CO_2 concentration in the ambient air [120]. Greater leaf thickness and more leaves per plant are noted when plants are grown in elevated CO_2 environments. Increased leaf growth results more often from increased cell expansion than from increased cell division. Crop plants exhibit greater leaf thickness than leaves from wild plant species when grown at increased CO_2 concentration [120].

6.6 Conclusions on the effect of climate change on fumonisin

Maize is commonly contaminated with fumonisin. Predicting climate change effects on fumonisin in maize is challenging, since the mycotoxin is an end product of the interaction between two living partners: the maize plant and the fungus parasitizing the host plant and synthesizing fumonisin in the maize kernels. The *Fusarium* spp. which produce fumonisin in maize are known in all continents where the crop is cultivated. In the temperate climates of Asia, Europe, and North America the risk for fumonisin contamination is highest at lower altitudes and latitudes. In tropical and subtropical countries the latitude is less important. The risk for fumonisin contamination is lower at high altitudes than in the lowlands of the tropical and subtropical countries of Africa, Asia, and South America.

The IPCC has published the opinion that the temperature increase will be highest over land and in northern latitudes. For the major maize producing countries of the Northern Hemisphere there is high confidence that the temperature increase will be in the range of 0.3–0.7 °C over the next twenty years [104]. In the major maize producing areas of the world, precipitation will not compensate for the temperature-driven increase in evaporation. Higher temperatures combined with greater fluctuations in precipitation will increase plant stress and predispose maize to infection with fumonisin-producing fungi.

The IPCC predicts that precipitation will increase at high latitudes, while in tropical and subtropical areas precipitation will decrease [104]. The prevalence of drought will increase with higher temperatures. The risk for summer drought will be greatest in mid-continental areas. Water deficit in the period from silking to kernel milky stage will make maize more susceptible to development of fumonisin in the maize kernels [104].

The predicted elevation of CO_2 concentrations in the atmosphere during the 21st century is unlikely to affect the risk for fumonisin contamination of maize.

The IPPC is of the opinion that climate change has negatively affected maize yields for many regions, and they state that there is medium confidence for this opinion [104]. Further climate change without adaptation is projected to further reduce maize production. The IPPC emphasizes that there is medium confidence in this statement [104].

The effects of climate change on plant diseases have not received much attention in the debate on the consequences of increased temperature and reduced precipitation on global crop production. Fungi producing fumonisin in maize have specific temperature and water availability requirements for infection and fumonisin production. Whether the reported increase in fumonisin contamination in dry and hot summers is the result of increased fungal growth in the maize kernels or of the up-regulation of *FUM* gene expression has not been elucidated.

However, both the currently observed changes in average climate, as well as the projected changes are relatively small compared to the variability in weather conditions which both the maize plant and the fumonisin producing fungi have to cope with during growing seasons, single crop seasons, and even sometimes within the diurnal cycle of a day. Although the key weather factors of air and soil temperature, rainfall, and air and soil moisture will change in range and possibly shift their patterns, the response of the plant and the pathogen to the relatively huge variability in current weather conditions provides ample opportunities to infer a robust causal relationship regarding the responses of fumonisin-producing fungi in maize under various weather conditions. We are somewhat surprised to observe that the scientific literature, including review articles, provides little emphasis on information from historical observations, compared to information derived from model simulations which typically take future climate scenarios as input.

The latter type of studies provides highly uncertain results, with the uncertainty resulting from both the climate scenarios and the scientific basis for fumonisin-producing fungi's responses to weather under various conditions. In general, the scientific literature reporting on studies on the potential impacts of climate change focus on scenarios for the average future weather conditions, which do not normally pretend to describe the full variability of future weather conditions. The latter component has, in studies of *Fusarium* head blight of wheat, been simulated with the additional aid of a weather generator [121]. However, the current and recent past variability in weather conditions actually completely covers the average future weather conditions of the next century, but not the full range of the expected future variability. These facts call for further studies that can reduce the uncertainty on how weather conditions affect the interaction of the maize plant and fumonisin-producing fungi.

References

[1] Sheldon JL. A corn mold. *Annu Rep Agric Exp Stn Nebr* 1904;17:23–32.

[2] Seifert KA, Aoki T, Baayen RP et al. The name *Fusarium moniliforme* should no longer be used. *Mycol Res* 2003;107:643–4.

[3] Marasas WFO, Wehner FC, van Rensburg SJ, van Schalkwyk DJ. Mycoflora of corn produced in human esophageal cancer areas in Transkei, Southern Africa. *Phytopathology* 1981;71: 792–796.

[4] Wilson TM, Nelson PF, Ryan TB. Linking leukoencephalomalacia to commercial horse rations. *Vet Med* 1985;80:63–69.

[5] Gelderblom WCA, Jaskiewicz J, Marasas WFO et al. Fumonisins: Novel mycotoxins with cancer-promoting activity produced by *Fusarium moniliforme. Appl Environ Microbiol* 1988;54:1806–1811.

[6] Bezuidenhout SG, Gelderblom WCA, Gorst-Allman CP et al. Structure elucidation of the fumonisins, mycotoxins from *Fusarium moniliforme. J Chem Soc Chem Commun* 1988;11: 743–745.

[7] Marasas WFO, Kellerman TS, Gelderblom WCA, Coetzer JA, Thiel PG, van der Lugt JJ. Leukoencephalomalacia in a horse induced by fumonisin B_1 isolated from *Fusarium moniliforme.* Onderstepoort *J Vet Res* 1988;55:197–203.

[8] Rheeder JP, Marasas WFO, Thiel PG, Sydenham EW, Shepard GS, van Schalkwyk DJ. *Fusarium moniliforme* and fumonisins in corn in relation to human esophageal cancer in Transkei. *Phytopathology* 1992;82:353–357.

[9] Harrison LR, Colvin BM, Greene JT, Newman LE, Cole JR. Pulmonary edema and hydrothorax in swine produced by fumonisin B_1, a toxic metabolite of *Fusarium moniliforme. J Vet Diagn Investig* 1990;2:217–221.

[10] Hendricks K. Fumonisins and neural tube defects in south Texas. *Epidemiology* 1999;10: 198–200.

[11] Marasas WFO. Fumonisins: Their implications for human and animal health. *Nat Toxins* 1995;3:193–198.

[12] Naik DM, Nawa IN, Raemakers RH. Absence of an effect from internally seed-borne Fusarium moniliforme on emergence, plant growth and yield of maize. *Seed Sci Technol* 1982;10: 347–356.

[13] Doehlert DC, Knutson CA, Vesonder RF. Phytotoxic effects of fumonisin B_1 on maize seedling growth. *Mycopathologia* 1994;127:117–121.

[14] Lamprecht SC, Marasas WFO, Alberts JF et al. Phytotoxicity of fumonisins and TA-toxin to corn and tomato. *Phytopathology* 1994;84:383–391.

[15] International Agency for Research on Cancer (IARC). Toxins derived from *Fusarium moniliforme*: fumonisins B_1 and B_2 and fusarin C. *In IARC Monographs on the evaluation of the carcinogenic risks to humans: Some naturally occurring substances: Food items and constituents, heterocyclic aromatic amines and mycotoxins.* Lyon, France: IARC; 1993;56: 445–466.

[16] Wang J, Zhou Y, Liu W, Zhu X, Du L, Wang Q. Fumonisin level in corn-based food and feed from Linxian County, a high-risk area for esophageal cancer in China. *Food Chemistry* 2008;106:241–246.
[17] Alizadeh AM, Roshandel G, Roudbarmohammadi S, et al. Fumonisin B_1 contamination of cereals and risk of esophageal cancer in high risk area in Northeastern Iran. *Asian Pacific J Cancer Prev* 2012;13:2625–2628.
[18] Wollenweber HW, Reinking OA. *Die Fusarien, ihre Beschreibung, Schadwirkung und Bekämpfung*. Berlin, Germany: Paul Parey; 1935.
[19] Leslie JF, Summerell BA. *The Fusarium laboratory manual*. Ames, Iowa, USA: Blackwell Publishing; 2006.
[20] Rheeder JP, Marasas WFO, Vismer HF. Production of fumonisin analogs by *Fusarium* species. *Appl Environ Microbiol* 2002;68:2101–2105.
[21] Bailly JD, Querin A, Tardieu D, Guerre P. Production and purification of fumonisins from a highly toxigenic *Fusarium verticillioides* strain. *Revue Méd Vét* 2005;156:547–554.
[22] Desjardins A. *Fusarium mycotoxins. Chemistry, genetics, and biology*. St. Paul, Minnesota, USA: American Phytopathological Society; 2006.
[23] Nelson PE, Plattner RD, Shackelford DD, Desjardins AE. Fumonisin B_1 production by *Fusarium* species other than *F. moniliforme* in Section *Liseola* and by some related species. *Appl Environ Microbiol* 1992;58:984–989.
[24] Frisvad JC, Smedsgaard J, Samson RA, Larsen TO, Thrane U. Fumonisin B_2 production by *Aspergillus niger*. *J Agric Food Chem* 2007;55:9727–9732.
[25] Frisvad JC, Larsen TO, Thrane U et al. Fumonisin and ochratoxin production in industrial *Aspergillus niger* strains. *PLoS ONE* 2011;6(8):e23496. doi:10.1371/journal.pone.0023496
[26] Soares C, Calado T, Venâncio A. Mycotoxin production by *Aspergillus niger* aggregate strains isolated from harvested maize in three Portuguese regions. *Rev Iberoam Micol* 2013;30:9–13.
[27] Nelson PE, Desjardins AE, Plattner RD. Fumonisins, mycotoxins produced by *Fusarium* species: Biology, chemistry, and significance. *Ann Rev Phytopathol* 1993;31:233–252.
[28] Munkvold GP, Desjardins AE. Fumonisins in maize. Can we reduce their occurrence? *Plant Dis* 1997;81:556–565.
[29] Binder EM, Tan LM, Chin LJ, Handl J, Richard J. Worldwide occurrence of mycotoxins in commodities, feeds and feed ingredients. *Anim Feed Sci Technol* 2007;137:265–282.
[30] Munkvold GP. Epidemiology of *Fusarium* diseases and their mycotoxins in maize ears. *Eur J Plant Pathol* 2003;109:705–713.
[31] Munkvold GP, Carlton WM. Influence of inoculation method on systemic *Fusarium moniliforme* infection of maize plants grown from infected seeds. *Plant Dis* 1997;81:211–216.
[32] Cotten TK, Munkvold GP. Survival of *Fusarium moniliforme, F. proliferatum*, and *F. subglutinans* in maize stalk residue. *Phytopathology* 1998;88:550–555.
[33] Foley DC. Systemic infection of corn by *Fusarium moniliforme*. *Phytopathology* 1962;52: 870–872.
[34] Munkvold, McGee DC, Carlton WM. Importance of different pathways for maize kernel infection by *Fusarium moniliforme*. *Phytopathology* 1997;87:209–217.
[35] Murillo-Williams A, Munkvold GP. Systemic infection by *Fusarium verticillioides* in maize plants grown under three temperature regimes. *Plant Dis* 2008;92:1695–1700.
[36] Yates IE, Hiett KL, Kapczynski DR et al. GUS transformation of the maize fungal endophyte *Fusarium moniliforme*. *Mycol Res* 1999;103:129–136.
[37] Reid LM, Nicol RW, Ouellet T et al. Interaction of *Fusarium graminearum* and *F. moniliforme* in maize ears: Disease progress, fungal biomass, and mycotoxin accumulation. *Phytopathology* 1999;89:1028–1037.

[38] Picot A, Hourcade-Marcolla D, Barreau C et al. Interactions between *Fusarium verticillioides* and *Fusarium graminearum* in maize ears and consequences for fungal development and mycotoxin accumulation. *Plant Pathol* 2012;61:140–151.

[39] Logrieco A, Mulè G, Moretti A, Bottalico A. Toxigenic *Fusarium* species and mycotoxins associated with maize ear rot in Europe. *Eur J Plant Pathol* 2002;108:597–609.

[40] Shelby RA, White DG, Bauske EM. Differential fumonisin production in maize hybrids. *Plant Dis* 1994;78:582–584.

[41] Desjardins AE, Plattner RD, Lu M, Claflin LE. Distribution of fumonisins in maize ears infected with strains of *Fusarium moniliforme* that differ in fumonisin production. *Plant Dis* 1998;82:953–958.

[42] Doko MB, Rapior S, Visconti A, Schjøth JE. Incidence and levels of fumonisin contamination in maize genotypes grown in Europe and Africa. *J Agric Food Chem.* 1995;43:429–434.

[43] Clements MJ, Kleinschmidt CE, Maragos CM, Pataky JK, White DG. Evaluation of inoculation techniques for Fusarium ear rot and fumonisin contamination of corn. *Plant Dis* 2003;87: 147–153.

[44] Bush BJ, Carson ML, Cubeta MA, Hagler WM, Payne GA. Infection and fumonisin production by *Fusarium verticillioides* in developing maize kernels. *Phytopathology* 2004;94:88–93.

[45] Uegaki R, Kobayashi H, Tohno M, Tsukiboshi T. Identification of mycotoxin-producing *Fusarium* spp. isolated from corn and the changes in concentration of fumonisin during the cultivation period. *Grassland Sci* 2012;58:121–126.

[46] Picot A, Barreau C, Pinson-Gadais L et al. The dent stage of maize kernels is the most conducive for fumonisin biosynthesis. *Appl Environ Microbiol* 2011;77:8382–8390.

[47] Seo J-A, Proctor RH, Plattner RD. Characterization of four clustered and coregulated genes associated with fumonisin biosynthesis in *Fusarium verticillioides*. *Fungal Genet Biol* 2001;34:155–165.

[48] Visentin I, Montis V, Döll K et al. Transcription of genes in the biosynthetic pathway for fumonisin mycotoxins is epigenetically and differentially regulated in the fungal maize pathogen *Fusarium verticillioides*. *Eucaryotic Cell* 2012;11:252–259.

[49] Proctor RH, Desjardins AE, Plattner RD, Hohn TM. A polyketide synthase gene required for biosynthesis of fumonisin mycotoxins in *Gibberella fujikuroi* mating population A. *Fungal Genet Biol* 1999;27:100–112.

[50] Proctor RH, Desjardins AE, McCormick SP, Plattner RD, Alexander NJ, Brown DW. Genetic analysis of the role of trichothecene and fumonisin mycotoxins in the virulence of *Fusarium*. *Eur J Plant Pathol* 2002;108:691–698.

[51] Desjardins AE, Plattner RD. Fumonisin B_1–Nonproducing strains of *Fusarium verticillioides* cause maize (*Zea mays*) ear infection and ear rot. *J Agric Food Chem* 2000;48:5773–5780.

[52] Witkowski JF, Wedberg JL, Steffey KL et al. Bt corn and European corn borer. Long-term success through resistance management. University of Minnesota Extension: *Crop News* 2013; 1–15.

[53] Munkvold GP, Hellmich RL, Showers WB. Reduced Fusarium ear rot and symptomless infection in kernels of maize genetically engineered for European corn borer resistance. *Phytopathology* 1997;87:1071–1077.

[54] Folcher L, Delos M, Marengue E et al. Lower mycotoxin levels in Bt maize grain. *Agron Sustain Dev* 2010;30:711–719.

[55] Bowers E, Hellmich R, Munkvold G. Vip3Aa and Cry1Ab proteins in maize reduce *Fusarium* ear rot and fumonisins by deterring kernel injury from multiple Lepidopteran pests. *World Mycotox J* 2013;6:127–135.

[56] Cardwell KF, Kling JG, Maziya-Dixon B, Bosque-Pérez NA. Interactions between *Fusarium verticillioides, Aspergillus flavus*, and insect infestation in four maize genotypes in lowland Africa. *Phytopathology* 2000;90:276–284.

[57] Leslie JF, Zeller KA, Lamprecht SC, Rheeder JP, Marasas WFO. Toxicity, pathogenicity, and genetic differentiation of five species of *Fusarium* from sorghum and millet. *Phytopathology* 2005;95:275–283.

[58] Palacios SA, Ramirez ML, Zalazar MC et al. Occurrence of *Fusarium* spp. in durum wheat grains. *J Agric Food Chem* 2011;59:12 264–12 269.

[59] Desjardins AE, Busman M, Proctor RH, Stessman R. Wheat kernel black point and fumonisin contamination by *Fusarium proliferatum*. *Food Addit Contam* 2007;24:1131–1137.

[60] Busman M, Desjardins AE, Proctor RH. Analysis of fumonisin contamination and the presence of *Fusarium* in wheat with kernel black point disease in the United States. *Food Addit Contam* 2012;29:1092–1100.

[61] Tansakul N, Limsuwan S, Trongvanichnam K. Fumonisin monitoring in Thai red cargo rice by reversed-phase high-performance liquid chromatography with electrospray ionization trap mass spectrometry. *Int Food Res J* 2012;19:1561–1566.

[62] Bansal J, Pantazopoulus P, Tam J et al. Surveys of rice sold in Canada for aflatoxins, ochratoxin A and fumonisins. *Food Addit Contam Part A* 2011;28:767–774.

[63] Noonim P, Mahakarnchanakul W, Nielsen KF, Frisvad JC, Samson RA. Fumonisin B_2 production by *Aspergillus niger* in Thai coffee beans. *Food Addit Contam Part A* 2009;26:94–100.

[64] Liu C, Xu W, Liu F, Jiang S. Fumonisin production by *Fusarium proliferatum* strains isolated from asparagus crown. *Mycopathologia* 2007;164:127–134.

[65] Marín S, Ramos AJ, Vázquez C, Sanchis V. Contamination of pine nuts by fumonisin produced by strains of *Fusarium proliferatum* isolated from *Pinus pinea*. *Lett Appl Microbiol* 2007;44:68–72.

[66] Stratakou I, van der Fels-Klerx HJ. Mycotoxins in grapes and wine in Europe: Occurrence, factors affecting the occurrence and related toxicological effects. *World Mycotox J* 2010;3: 283–300.

[67] Logrieco A, Ferracane R, Visconti A, Ritieni A. Natural occurrence of fumonisin B_2 in red wine from Italy. *Food Addit Contam Part A* 2010;27:1136–1141.

[68] Mogensen JM, Larsen TO, Nielsen KF. Widespread occurrence of the mycotoxin fumonisin B_2 in wine. *J Agric Food Chem* 2010;58:4853–4857.

[69] Cozzi G, Paciolla C, Haidukowski M, Leonardis S de, Mule G, Logrieco A. Increase of fumonisin B_2 and ochratoxin A production by black *Aspergillus* species and oxidative stress in grape berries damaged by powdery mildew. *J Food Prot* 2013;76:2031–2036.

[70] Chiotta ML, Susca A, Stea G et al. Phylogenetic characterization and ochratoxin A – Fumonisin profile of black *Aspergillus* isolated from grapes in Argentina. *Int J Food Microbiol* 2011;149:171–176.

[71] Hlywka JJ, Bullerman LB. Occurrence of fumonisin B_1 and B_2 in beer. *Food Addit Contam* 1999;16:319–324.

[72] Shephard GS, Van der Westhuizen L, Gatyeni PA, Smodyala NIM, Burger H-M, Marasas WFO. Fumonisin mycotoxins in traditional Xylosa maize beer in South Africa. *J Agric Food Chem* 2005;53:9634–9637.

[73] Pietri A, Bertuzzi T, Agosti B, Donadini G. Transfer of aflatoxin B_1 and fumonisin B_1 from naturally contaminated raw materials to beer during an industrial brewing process. *Food Addit Contam Part A* 2010;27:1431–1439.

[74] Fandohan P, Hell K, Marasas WFO, Wingfield MJ. Infection of maize by *Fusarium* species and contamination with fumonisin in Africa. *African J Biotech* 2003;2:570–579.

[75] Bankole S, Schollenberger M, Drochner W. Mycotoxins in food systems in Sub Saharan Africa: A review. *Mycotoxin Res* 2006;22:163–169.
[76] Bankole SA, Mabekoje OO. Occurrence of aflatoxins and fumonisins in preharvest maize from south-western Nigeria. *Food Addit Contam* 2004;21:251–255.
[77] Nikièma PN, Worrillow L, Traoré AS, Wild CP, Turner PC. Fumonisin contamination of maize from Burkina Faso, West Africa. *Food Addit Contam* 2004;21:865–870.
[78] Fandohan P, Gnonlonfin G, Hell K, Marasas WFO, Wingfield MJ. Natural occurrence of *Fusarium* and subsequent fumonisin contamination in preharvest and stored maize in Benin, West Africa. *Int J Food Microbiol.* 2005;99:173–183.
[79] Sangare-Tigori B, Moukha S, Kouadio HJ, Betbeder A-M, Dano DS, Creppy EE. Co-occurrence of aflatoxin B_1, fumonisin B_1, ochratoxin A and zearalenone in cereals and peanuts from Côte d'Ivoire. *Food Addit Contam* 2006;23:1000–1007.
[80] Ayalew A, Fehrmann H, Lepschy J, Beck R, Abate D. Natural occurrence of mycotoxins in staple cereals from Ethiopia. *Mycopathologia* 2006;162:57–63.
[81] Atukwase A, Kaaya AN, Muyanja C. Factors associated with fumonisin contamination of maize in Uganda. *J Sci Food Agric* 2009;89:2393–2398.
[82] Atukwase A, Kaaya AN, Muyanja C. Dynamics of *Fusarium* and fumonisins in maize during storage – A case of traditional storage structures commonly used in Uganda. *Food Control* 2012;26:200–205.
[83] Alkonya AE, Monda EO, Ajanga S. Variation in *in vitro* fumonisin B_1 production by different *Fusarium verticillioides* isolates in Kenya. *American-Eurasian J Agric & Environ Sci* 2008;4:368–371.
[84] Schjøth JE, Visconti A, Sundheim L. Fumonisins in maize in relation to climate, planting time and hybrids in two agroecological zones in Zambia. *Mycopathologia* 2009;167:209–219.
[85] Mukanga M, Derera J, Tongoona P, Laing MD. A survey of pre-harvest ear rot diseases of maize and associated mycotoxins in south and central Zambia. *Int J Food Microbiol* 2010;141:213–221.
[86] Nooh A, Amra H, Youssef MM, El-Banna AA. Mycotoxin and toxigenic fungi occurrence in Egyptian maize. *Int J Adv Res* 2014;2:521–532.
[87] European Commission 2007. Commission Regulation (EC) No 1126/2007 of 28.9.2007 amending regulation (EC) No 1181/2006 setting maximum levels for certain contaminants in foodstuffs as regards *Fusarium* toxins in maize and maize products. *Off J Eur Union L* 2007;254:14–17.
[88] Visconti A, Doko MB. Survey of fumonisin production by *Fusarium* isolated from cereals in Europe. *J AOAC Int* 1994;77:546–550.
[89] Ariño A, Herrera A, Juan T et al. Influence of agricultural practices on the contamination of maize by fumonisin mycotoxins. *J Food Prot* 2009;72:898–902.
[90] Ivic D, Kovacevik B, Vasilj V, Idzakovic N. Occurrence of potentially toxigenic *Fusarium verticillioides* and low fumonisin B_1 content on barley grain in Bosnia and Herzegovina. *J Appl Bot Food Quality* 2011;84:121–124.
[91] Van Asselt ED, Azambuja W, Moretti A et al. A Dutch field survey on fungal infection and mycotoxin concentration in maize. *Food Addit Contam Part A.* 2012;29:1556–1565.
[92] Nielsen LK, Jensen JD, Nielsen GC et al. *Fusarium* head blight of cereals in Denmark: Species complex and related mycotoxins. *Phytopathology* 2011;101:960–969.
[93] Parikka P, Hakala K, Tiilikkala K. Expected shifts in *Fusarium* species' composition on cereal grain in Northern Europe due to climatic change. *Food Addit Contam Part A* 2012;29: 1543–1555.
[94] Giannitti F, Diab SS, Pacin AM et al. Equine leukoencephalomalacia (ELEM) due to fumonisins B1 and B2 in Argentina. *Pesq Vet Bras* 2011;31:407–412.

[95] Broggi LE, Pacin AM, Gasparovic A et al. Natural occurrence of aflatoxins, deoxynivalenol, fumonisins and zearalenone in maize from Entre Rios Province, Argentina. *Mycotoxin Res* 2007;23:59–64.
[96] Ono EYS, Ono MA, Funo FY et al. Evaluation of fumonisin-aflatoxin co-occurrence in Brazilian corn hybrids by ELISA. *Food Addit Contam Part A* 2001;18:719–729.
[97] Stumpf R, dos Santos J, Gomes LB et al. Fusarium species and fumonisins associated with maize kernels produced in Rio Grande do Sul State for the 2008/09 and 2009/10 growing seasons. *Braz J Microbiol* 2013;44:89–95.
[98] Hooker DC, Schaafsma AW. Agronomic and environmental impacts on concentration of deoxynivalenol and fumonisin B_1 in corn across Ontario. *Can J Plant Pathol* 2005;27:347–356.
[99] Jurjevic Z, Wilson DM, Wilson JP et al. *Fusarium* species of the *Gibberella fujikuroi* complex and fumonisin contamination of pearl millet and corn in Georgia, USA. *Mycopathologia* 2005;159:401–406.
[100] Wu F, Bhatnager D, Bui-Klimke T et al. Climate change impacts on mycotoxin risks in US maize. *World Mycotox J* 2011;4:79–93.
[101] Wei T, Zhu W, Pang M, Liu Y, Dong JG. Natural occurrence of fumonisins B_1 and B_2 in corn in four provinces of China. *Food Addit Contam Part B* 2013;6:270–274.
[102] Azizi IG, Rouhi S. The comparison of total fumonisin and total aflatoxin levels in biscuit and cookie samples in Babol City, Northern Iran. *Iranian J Publ Health* 2013;42:422–427.
[103] Trung TS, Tabuc C, Bailly S, Querin A, Guerre P, Bailly JD. Fungal mycoflora and contamination of maize from Vietnam with aflatoxin B_1 and fumonisin B_1. *World Mycotox J* 2008;1:87–94.
[104] Stocker TF, Qin D, Plattner G-K, Tignor M, Allen SK, Boschung J, Nauels A, Xia Y, Bex V, Midgley PM (eds). IPCC 2013: Summary for Policymakers. In: *Climate Change 2013: The Physical Science Basis. Contribution of Working Group I to the Fifth Assessment Report of the Intergovernmental Panel on Climate Change*. Cambridge, United Kingdom and New York, NY, USA: Cambridge University Press; 2013.
[105] Oerke E-C. Crop losses to pests. *J Agric Sci* 2006;144:31–43.
[106] Coakley SM, Scherm H, Chakraborty S. Climate change and plant disease management. *Ann Rev Phytopathol* 1999;37:399–426.
[107] de la Campa R, Hooker DC, Miller JD, Schaafsma AW, Hammond BG. Modelling effects of environment, insect damage, and *Bt* genotypes on fumonisin accumulation in maize in Argentina and the Philippines. *Mycopathologia* 2005;159:539–552.
[108] Schaafsma AW, Hooker DC. Climatic models to predict occurrence of *Fusarium* toxins in wheat and maize. *Int J Food Microbiol* 2007;119:116–125.
[109] Marin S, Sanchis V, Magan N. Water activity, temperature and pH effects on growth of *Fusarium moniliforme* and *Fusarium proliferatum* isolates from maize. *Can J Microbiol* 1995;41:1063–1070.
[110] Marin S, Sanchis V, Vinas I, Canela R, Magan N. Effect of water activity and temperature on growth and fumonisin B_1 and B_2 production by *Fusarium proliferatum* and *F. moniliforme* on maize grain. *Letters Appl Microbiol* 1995;21:298–301.
[111] Cao A, Santiago R, Ramos AJ et al. Critical environmental and genotypic factors for *Fusarium verticillioides* infection, fungal growth and fumonisin contamination in maize grown in northwestern Spain. *Int J Food Microbiol* 2014;177:63–71.
[112] Jurado M, Marín P, Magan N, González-Jeán MT. Relationship between solute and matric potential stress, temperature, growth and *FUM1* gene expression in two *Fusarium verticillioides* strains from Spain. *Appl Environ Microbiol* 2008;74:2032–2036.
[113] Marín P, Magan N, Vázquez C, González-Jeán MT. Differential effect of environmental conditions on the growth and regulation of the fumonisin biosynthesis gene *FUM1* in the maize

pathogens and fumonisin-producers *Fusarium verticillioides* and *Fusarium proliferatum*. *FEMS Microbiol Ecol* 2010;73:303–311.

[114] Parsons MW, Munkvold GP. Effects of planting date and environmental factors on fusarium ear symptoms and fumonisin B_1 accumulation in maize grown in six North American locations. *Plant Pathol* 2012;61:1130–1142.

[115] Elsgaard L, Børgesen CD, Olesen JE et al. Shifts in comparative advantages for maize, oat and wheat cropping under climate change in Europe. *Food Addit Contam Part A* 2012;29:1514–1526.

[116] Camargos SM, Soares LMV, Sawazaki E, Bolonhezi D, Castro JL, Bortolleto N. Accumulation of fumonisin B_1 and B_2 in freshly harvested Brazilian commercial maize at three locations during two nonconsecutive seasons. *Mycopathologia* 2001;155:219–228.

[117] Ono EYS, Sugiura Y, Homechin M et al. Effect of climatic conditions on natural mycoflora and fumonisins in freshly harvested corn of the State of Paraná, Brazil. *Mycopathologia* 1999;147:139–148.

[118] Herrera M, Conchello P, Juan T, Estopañan G, Herrera A, Ariño A. Fumonisin concentrations in maize as affected by physio-chemical, environmental and agronomic conditions. *Maydica* 2010;55:121–126.

[119] Magan N, Medina A, Aldred D. Possible climate-change effects on mycotoxin contamination of food crops pre- and postharvest. *Plant Pathol* 2011;60:150–163.

[120] Pritchard SG, Rogers HH, Prior SA, Peterson CM. Elevated CO_2 and plant structure: A review. *Global Change Biol* 1999;5:807–837.

[121] Semenow MA, Pilkington-Bennett S, Calanca P. Validation of ELPIS 1980–2010 baseline scenarios using the observed European climate assessment data set. *Clim Res* 2013;57:1–9.

Maria Paula Kovalsky Paris, Yin-Jung Liu, Karin Nahrer, and
Eva Maria Binder

7 Climate change impacts on mycotoxin production

7.1 Introduction

Prices of staple grains have increased significantly in the last five years and climate change is bound to add pressure to the supply and quality of food, and the sustainability of production practices worldwide. The impact of climate change on food production has been identified as an emerging hazard for food and feed security worldwide [1, 2], and the consequent risks are being received with increasing scientific attention [3]. One key aspect in food and feed safety is the changing pattern of mycotoxin contamination in cereals such as wheat, maize, and rice due to climate change, which the EFSA's Emerging Risks Unit has recognized as a potential emerging hazard [1].

The 2014 Intergovernmental Panel on Climate Change (IPCC) report illustrates different global warming projections depending on low- or high-emission scenarios. The prediction for the year 2100 is that global temperatures may increase by up to 4.8 °C [4]. The world is undergoing extreme weather events, such as hurricanes, which are causing the spread of plant pathogens to new areas [5]. This is expected to increase with the projected increase in the frequency of extreme weather events due to climate change [6].

The distribution of species around the world will also show an overall shift polewards. Bebber *et al.* found that plant pathogens and pests are moving at a rate of about 2.7 km/year towards the poles [7], which is very close to the rate of climate change [8]. For some areas, projections of decreasing summer precipitation and increases in temperature would ultimately lead to drought stress episodes. Changes are also expected to occur among crops grown in the next 10 to 20 years due to the predicted rise in atmospheric CO_2 concentrations at a rate of 1.5 µmol/year [3].

The impact of climate change on fungal colonization has not yet been fully elucidated. However, it is known that temperature, humidity, and precipitation have an effect on toxigenic fungi and on pathogen-plant host interaction [9]. It has also been suggested that slightly elevated CO_2 concentrations and interactions with temperature and water availability may stimulate the growth of some mycotoxigenic fungi, particularly those under water stress [3]. Displacements of fungal species by other more virulent or aggressive fungi have already been observed [10]. Other secondary factors may have an additional impact on the mycotoxigenic fungal profile, including insect attacks and effectiveness of fungicides and pesticides, as well as the alteration of geographical distribution or the life cycle of insects which promote fungal infections in crops [3].

7.2 Impact of temperature, water availability, and CO_2 on mycotoxin production

The life cycle of all microorganisms, including mycotoxigenic molds, depends on two main factors: water availability and temperature [11]. These factors interact to influence a series of parameters including germination, growth, sporulation, and mycotoxin production [12]. In general, there is a temperature range within which fungi perform best. Therefore, increasing average temperatures may alter the range at which fungi are able to compete.

As concentrations of methane, carbon dioxide, nitrous oxide, and chlorofluorocarbons in the atmosphere increase, so will incidences of environmental warming, higher precipitation, or drought [13]. These extreme weather conditions will have wide-ranging impacts, depending on world region and mycotoxin systems [14]. Several authors have reviewed some aspects of the impact that climate change may have on plant breeding, plant diseases, and mycotoxins in Europe, Australia, and Africa [2, 15–18]. There are some predicted effects of temperature increase and water availability on both fungal growth rates and on the production of some of the most relevant mycotoxins: *Alternaria* toxins, fumonisins, trichothecenes, ochratoxin A, aflatoxin B_1, and patulin. However, these predictions do not take the possible additional effect of CO_2 into account [3]. Typically, high concentrations of mycotoxins in grains depend on the number of rainy days and days with relative humidity above 75 %, but there can be a decrease in mycotoxin concentrations at temperatures below 12 or above 32 °C [19].

Aspergillus flavus and *A. parasiticus*, the main aflatoxin producers, are xerophilic fungi which thrive under higher temperatures and lower rainfall [1]. For instance, during the hot and dry episodes in northern Italy in 2003, *A. flavus* was able to actively colonize the ripening maize, a key crop, by outcompeting the more common *Fusarium* species [20]. This led to an uncommon increase in Afla B_1 contamination in Europe.

Fumonisin occurrence is linked to drought stress. Dry season maize in Southern and Eastern Africa may contain high levels of this toxin, even if the maize appears to be of very good quality. Fumonsins are less significant in northern temperate zones with cooler climates. High temperatures favor the growth of the fumonisin producer *Fusarium verticillioides*, which means that a warming trend would lead to greater domination of the fungus compared to other maize-borne *Fusarium* species [9].

The displacement of the formerly predominant species, *F. culmorum* and *Microdochium nivale*, by the more virulent plant pathogen *F. graminearum* as a result of warm European summers has been reported [9]. Since *M. nivale* is nontoxigenic and *F. culmorum* generally produces a lower number of mycotoxins than *F. graminearum*, the concentration of mycotoxins may consequently increase [6]. In Canada, a more toxigenic 3-acetylated deoxynivalenol (3ADON) chemotype of *F. graminearum* replaced the 15-acetylated deoxynivalenol (15ADON) chemotype, indicating genetic differentiation [21]. Novel strains which form unexpected toxins are also being discovered and may have resulted from changing environments. Such shifts in mycotoxigenic fungi may also lead to changes in the mycotoxin chemical profile [9]. Minnesota, for instance, has witnessed the emergence of a novel *Fusarium* isolate called the "Northland population" which does not produce the trichothecenes deoxynivalenol or nivalenol [10].

Climate change is expected to increase the biomass of crops and alternative host plants will further increase inoculum production. Chakraborty *et al.* have shown that an increase in atmospheric CO_2 concentration will directly increase the amounts of Fusarium head blight (FHB) and Crown Rot (CR) inoculum. The authors showed that the saprotrophic fitness of *F. graminearum* remained at elevated CO_2 levels and did not suffer any decrease in its ecological fitness [6].

A deterioration of grain quality may occur as a direct effect of increasing temperature and CO_2, resulting in a reduction in the protein and micronutrient content in grain. Such an effect could facilitate mold growth and mycotoxin production, which could affect quality during storage and transport [6]. It has been shown that extreme drought episodes, desertification, and variations in wet/dry cycles will impact mycotoxigenic fungi. In developing countries, drought stress may play an important role in terms of food security. Maize, peanuts, and pistachios are particularly prone to infection during heat/drought stress due to cracking or splitting, which may result in a significant increase in *A. flavus* and, consequently, aflatoxin contamination. An increase in preharvest aflatoxin contamination will affect nutritional quality and have an impact on consumption or the ability to export [9].

Water availability is predicted to be altered by climate change, and some areas will experience drought while others experience greater precipitation. The effects of humidity on mycotoxin production in crops are less straightforward than those for temperature [14]. In a postharvest scenario, grain silos which harbor pests which can multiply more rapidly under warmer temperatures could contain higher amounts of metabolic water. This increase in condensation could initiate spoilage, potentially increasing contamination with mycotoxins such as ochratoxin A, aflatoxins, and perhaps trichothecenes in damp grain. This means that the distribution and types of mycotoxins postharvest may change significantly, giving way to new emerging mycotoxins [9].

7.3 Prediction strategies

Geostatistics is a tool which enables the investigation of specific regional hotspots which may represent high risk of mycotoxin contamination based on meteorological data. This method could be used to predict the optimal time points for fungicide applications and pest control in order to minimize contamination [3].

Aflatoxin prediction strategies have already been applied worldwide. For instance, Battilani *et al.* studied the regions in Europe which may be hot spots for

A. carbonarius contamination of wine grapes and at a higher risk for contamination with ochratoxin A [22]. The authors showed that the highest contamination with this fungus was just prior to harvest. In Australia, recent studies have demonstrated that it is possible to use an agricultural production systems simulator to calculate an aflatoxin risk index (ARI) in both maize and peanuts [17]. The use of advanced high resolution radiometer (AVHRR) satellite data has also been used in Mali, West Africa, to examine the relationship between peanut yield, total precipitation, and the maximum reproductive phase of peanut plants and aflatoxin contamination [16].

The main prediction tool for Fusarium head blight (FHB) and the risk of DON contamination in small grain cereals and maize is DONCAST, a model based on the weather conditions just prior to and during wheat head emergence. The method includes the input of local or regional weather data, the number of rainy days prior to and after anthesis. Other factors which are taken into account include weather data during ripening, and maximum, minimum and mean temperatures of the region. Other approaches to DON prediction are based on the life cycle of *F. graminearum* in a method called the TOX-risk, which is a risk index calculated daily for *F. graminearum* and *F. culmorum* over the growing season [23, 24].

A similar approach to that used for DONCAST was developed to predict fumonisin contamination, taking climatic information relevant to the growing period into account. Changes in temperature and the amount of rain post-silking were used to predict the risk of fumonisins in maize and genetically modified maize [19, 25].

The potential use of the geographical emerging mycotoxin identification system (GEMIS) model has been tested across Europe to examine the effects of changes in temperature (+2 °C) and rainfall (+3 mm) on heading date and cereal productivity areas [26].

7.4 Other factors to consider

Temperature, water availability, and CO_2 are believed to be the most important components affecting changes in mycotoxin patterns. However, other less considered climate-influenced factors such as insect and other pest attacks, the effectiveness of fungicides, and the possible effects on mycotoxin biosynthetic pathways should also be recognized as potential and indirect triggers of fungi colonization and mycotoxin production [9].

Insect attacks are known to lower the resistance of plants to stress, and the resulting wounds on kernels may favor infection by the fungus. There is substantial evidence to indicate that there will be an overall increase in the number of outbreaks of various insects [9]. The IPCC report predicts that more insect pests will occur at higher temperatures [4], leading to a higher distribution of mycotoxigenic fungi [27, 28]. Also, more insects may increase insect-feeding birds which may result in more bird damage to crops, making the ecological implications of climate change even more complex [14].

Fungicide effectiveness may be significantly modified by climate variation, leading to less effective control of mycotoxigenic pathogens and toxin contamination [3]. Medina *et al.* showed that both water availability and temperature modulate the effect of carbendazim [29] and natamycin [30] against ochratoxin A-producing *A. carbonarius*.

Mycotoxin biosynthetic pathways have been elucidated for many of the key mycotoxigenic fungi, such as the aflatoxin B_1-producing *A. flavus*, the DON-producing *F. graminearum*, and the fumonisin B_1-producing *F. verticillioides*. Genes involved in the production of mycotoxins are often clustered together. Microarray analysis has been used to investigate the effect of stress factors on such genes. Schmidt-Heydt *et al.* suggested that changes in both temperature and water availability may increase mycotoxin production [31, 32].

7.5 Insights into potential mycotoxin production: focus on Europe

Patterson *et al.* have suggested that just as in the scenario where climate change will lead to an increase in total mycotoxins because more crops are produced within a region, conversely there would be a decrease in total mycotoxins because fewer crops are produced within a region [14]. According to the IPCC report, more crops or higher yields will occur in regions which are currently cool, and fewer crops or yields will occur in the currently warm regions [4]. However, warm regions are expected to see more encroachment of arid and semi-arid land, heightening drought stress in crops and leading perhaps to an increase in mycotoxins. In currently cool regions, storage conditions may worsen as temperatures increase [14].

According to the IPCC, it is predicted that annual precipitation in the Mediterranean regions of Europe and Africa will decrease [4]. An increase in aflatoxins and ochratoxin A from *Aspergillus* spp., as well as fumonisins is expected in sub-Mediterranean countries as a result of temperature increases. For instance in northern Portugal, the ochratoxin A producer *A. carbonarius* was the main mycotoxigenic fungus found on grapes [33, 34]. *A. flavus* and *Penicillium expansum* were present in lower amounts. It is predicted that in 100 years, *A. flavus* may outcompete *A. carbonarius*, increasing the threat of aflatoxins compared to ochratoxin A. Temperatures may also be too high for *P. expansum*, resulting in a possible reduction of patulin [14]. Aflatoxins have been reported in northern Italy [20], while *A. flavus* has been isolated in Hungary in regions where the fungus was not present before [35].

High temperatures and drought are expected to reduce water availability and consequently crop productivity in the south of Europe. Daily precipitation is expected to increase in northern Europe [4]. In southern and south-eastern Europe, including Portugal, Spain, southern France, Italy, Slovenia, Greece, Malta, Cyprus, Bulgaria, and southern Romania, an increase of 4–5 °C is predicted and water availability is expected to be reduced, especially in the summer [14].

Western and Atlantic European countries, such as Benelux, western and northern France, northern Germany, United Kingdom, Irish Republic, the Netherlands, and Denmark are predicted to experience an increase of 2.5–3.5 °C with drier and hotter summers [14].

7.6 Trends in mycotoxin occurrence

As part of its mycotoxin risk management program, Biomin has been conducting a yearly mycotoxin survey since 2004 which provides insight into the risks caused by the main mycotoxins found in agricultural commodities such as corn, wheat, barley, silage, as well as finished feed and others. Till now, more than 26 000 samples from different countries around the world have been evaluated for the presence of aflatoxins (Afla), zearalenone (ZEN), deoxynivalenol (DON), fumonisins (FUM) and ochratoxin A (OTA). The Biomin Mycotoxin Survey summarizes the importance of the co-occurrence of various mycotoxins in different samples.

Table 7.1 provides a global overview of the mycotoxins present in all commodities analyzed each year. The percentage of samples above the EC guideline for aflatoxin B_1 or the EC recommended levels for all other toxins are indicated in the last column of the table to allow comparison of the results. These percentages are illustrated to provide an overview of the trends for the particular mycotoxin across the different years. The results worldwide depict the changes in the most problematic mycotoxins for each year. For instance, zearalenone levels were particularly high in 2006; the highest ochratoxin contaminations were observed in 2007; the contamination of aflatoxins was at its highest so far in 2008; fumonisin levels were at a maximum in 2009; and 2010 is considered a DON-contamination peak year.

As DON is one of most commonly occurring mycotoxins worldwide, it was the most extensively analyzed within the mycotoxin survey. In 2005, a total of 1432 samples were analyzed for DON content and the number of samples increased each year to almost 4000 samples in 2013. The EC recommended value for complementary and complete feedstuffs for pigs (900 ppb) was used for means of comparison [36]. The percentage of samples over the years with concentrations of DON above 900 ppb show fluctuations between a minimum of 9.4 % in the first half of 2014 and a maximum of 18.2 % in 2010.

The global occurrence of zearalenone was also intensively studied within the mycotoxin survey. The EC recommendation for ZEN in complementary and complete feedstuffs for piglets and gilts (young sows) is set at 100 ppb [36], due to the high sensitivity of these animals to this estrogen-like substance. There was a high variation in the percentage of samples above 100 ppb, a level which lies above the EC recommendation. This ranged from 9.5 % in 2013 to 22.4 % of all samples in 2006, which is a considerably large number of samples that could potentially affect sow fertility.

From 2005 to 2013, a total of 700 to 2800 samples were analyzed in the respective years for total aflatoxin content. Results show that 13.2 % (in 2012) to almost 24 % (in 2008) of the samples analyzed lie above the European maximum value of 5 ppb for aflatoxin B_1 for complete feedstuffs for dairy animals [37]. It is well known that afla-

Tab. 7.1: Results of the BIOMIN Mycotoxin Survey for all commodities analyzed worldwide.

Mycotoxin	Year	# samples	100–299 ppb	300–899 ppb	900–999 ppb	1000–1999 ppb	2000–4999 ppb	5000–7999 ppb	≥ 8000 ppb	≥ 900 ppb DON (EC recomm. pigs)
DON	2005	1432	10.3 %	21.2 %	1.2 %	6.8 %	2.9 %	0.2 %	0.1 %	11.2 %
	2006	1880	11.5 %	22.4 %	2.0 %	8.4 %	4.5 %	0.7 %	0.3 %	15.9 %
	2007	1289	15.1 %	18.3 %	1.6 %	6.2 %	3.2 %	0.8 %	1.2 %	13.0 %
	2008	1729	15.2 %	27.2 %	2.0 %	8.8 %	5.1 %	0.8 %	0.2 %	16.9 %
	2009	2413	12.5 %	23.4 %	1.5 %	6.7 %	3.3 %	0.4 %	0.3 %	12.2 %
	2010	**2948**	**12.8 %**	**23.8 %**	**1.7 %**	**8.4 %**	**5.7 %**	**1.6 %**	**0.8 %**	**18.2 %**
	2011	3509	15.0 %	23.1 %	1.6 %	8.1 %	5.3 %	0.9 %	0.7 %	16.6 %
	2012	3712	19.7 %	17.9 %	1.8 %	6.6 %	6.2 %	1.5 %	1.2 %	17.2 %
	2013	3931	18.5 %	21.3 %	1.1 %	5.9 %	4.2 %	0.5 %	0.5 %	12.2 %
	2014	1655	19.2 %	22.0 %	1.5 %	4.6 %	2.7 %	0.5 %	0.1 %	9.4 %

Mycotoxin	Year	# samples	20–49 ppb	50–99 ppb	100–199 ppb	200–249 ppb	250–499 ppb	500–1999 ppb	≥ 2000 ppb	≥ 100 ppb ZEN (EC recomm. pigs)
ZEN	2005	1110	4.8 %	8.6 %	6.5 %	2.9 %	4.3 %	2.8 %	1.1 %	17.6 %
	2006	**1663**	**4.8 %**	**7.9 %**	**8.2 %**	**2.3 %**	**4.8 %**	**5.4 %**	**1.7 %**	**22.4 %**
	2007	1121	12.6 %	9.8 %	5.3 %	1.2 %	2.2 %	1.2 %	0.8 %	10.7 %
	2008	1424	9.7 %	14.5 %	9.1 %	1.5 %	3.1 %	3.3 %	0.6 %	17.6 %
	2009	2303	8.4 %	9.6 %	6.2 %	1.4 %	3.5 %	2.2 %	0.5 %	13.8 %
	2010	2634	10.5 %	9.3 %	8.0 %	1.9 %	4.8 %	4.4 %	0.6 %	19.7 %
	2011	3061	11.6 %	8.9 %	7.0 %	1.8 %	4.0 %	3.4 %	0.4 %	16.6 %
	2012	3320	10.8 %	5.5 %	4.3 %	1.3 %	3.4 %	4.9 %	1.2 %	15.1 %
	2013	3470	11.8 %	6.5 %	4.9 %	0.9 %	1.8 %	1.6 %	0.3 %	9.5 %
	2014	1613	11.4 %	10.9 %	4.3 %	1.0 %	2.9 %	2.5 %	0.2 %	11.0 %

Mycotoxin	Year	# samples	1–4.9 ppb	5–9.9 ppb	10–19.9 ppb	20–199 ppb	≥ 200 ppb	≥ 5 ppb AfB_1 (EC max for dairy cattle)
Total aflatoxin B_1, B_2, G_1, G_2	2005	698	0.9 %	5.3 %	4.7 %	4.4 %	0.7 %	15.2 %
	2006	1132	2.2 %	5.2 %	4.4 %	8.2 %	0.8 %	18.6 %
	2007	807	2.9 %	6.2 %	5.7 %	8.2 %	1.2 %	21.3 %
	2008	**1132**	**7.1 %**	**4.7 %**	**4.9 %**	**12.1 %**	**2.1 %**	**23.8 %**
	2009	1661	9.7 %	4.8 %	4.9 %	10.5 %	2.8 %	22.9 %
	2010	1950	10.7 %	4.2 %	3.2 %	9.0 %	0.9 %	17.3 %
	2011	2770	10.6 %	2.6 %	3.0 %	8.2 %	1.9 %	15.7 %
	2012	2636	10.7 %	4.7 %	3.2 %	4.7 %	0.6 %	13.2 %
	2013	2839	12.8 %	5.6 %	3.7 %	5.5 %	0.8 %	15.6 %
	2014	1267	9.6 %	4.1 %	2.9 %	8.2 %	1.9 %	17.1 %

Tab. 7.1 (continued)

Mycotoxin	Year	# samples	200–999	1000–1499	1500–2999	3000–4999	5000–19 999	≥ 20 000		≥ 5000 ppb total fumonisins (EC recomm. pigs/horses)
			ppb							
Total fumonisin B_1, B_2	2005	698	30.2 %	3.3 %	4.3 %	1.0 %	1.1 %	0.3 %		1.4 %
	2006	1000	26.1 %	5.0 %	5.3 %	2.7 %	3.0 %	0.5 %		3.5 %
	2007	646	34.1 %	6.2 %	8.4 %	2.9 %	1.1 %	0.2 %		1.2 %
	2008	1074	32.4 %	5.7 %	5.8 %	2.8 %	2.7 %	0.0 %		2.7 %
	2009	**1579**	**22.1 %**	**7.3 %**	**9.4 %**	**5.3 %**	**6.6 %**	**0.4 %**		**7.1 %**
	2010	2108	23.2 %	7.2 %	9.6 %	4.6 %	4.3 %	0.2 %		4.5 %
	2011	2548	21.6 %	5.8 %	9.5 %	3.7 %	2.9 %	0.3 %		3.2 %
	2012	2570	26.1 %	6.8 %	8.9 %	3.2 %	2.3 %	0.1 %		2.5 %
	2013	2699	24.1 %	5.7 %	9.1 %	3.6 %	2.8 %	0.1 %		2.9 %
	2014	1298	27.7 %	5.9 %	8.5 %	4.2 %	3.4 %	0.3 %		3.7 %

Mycotoxin	Year	# samples	5–9.9	10–49	50–79	80–99	100–149	150–399	≥ 400	≥ 50 ppb OTA (EC recomm. pigs)
			ppb							
OTA	2005	145	2.8 %	2.1 %	0.0 %	0.0 %	0.7 %	1.4 %	0.7 %	2.8 %
	2006	269	4.5 %	5.6 %	1.1 %	0.4 %	0.0 %	0.4 %	0.0 %	1.9 %
	2007	**207**	**2.9 %**	**2.4 %**	**0.0 %**	**1.4 %**	**1.0 %**	**1.9 %**	**2.9 %**	**7.2 %**
	2008	930	4.1 %	3.5 %	0.5 %	0.1 %	0.2 %	0.1 %	0.0 %	1.0 %
	2009	1034	4.0 %	4.1 %	0.3 %	0.0 %	0.0 %	0.4 %	0.1 %	0.8 %
	2010	1558	4.4 %	2.4 %	0.4 %	0.1 %	0.1 %	0.1 %	0.0 %	0.7 %
	2011	1966	3.2 %	4.4 %	0.2 %	0.1 %	0.1 %	0.2 %	0.1 %	0.6 %
	2012	2230	3.6 %	2.8 %	0.4 %	0.0 %	0.0 %	0.0 %	0.0 %	0.5 %
	2013	2459	2.0 %	2.8 %	0.2 %	0.0 %	0.1 %	0.3 %	0.0 %	0.7 %
	2014	1113	1.8 %	1.9 %	0.4 %	0.0 %	0.0 %	0.0 %	0.0 %	0.4 %

toxins often co-occur with fumonisins. The animals most sensitive to fumonisins are pigs and horses, therefore the EC recommended level for fumonisins in complementary and complete feedstuffs for these animals has been set at 5000 ppb. A fumonisin contamination peak was observed in 2009, as 7.1 % of samples contained fumonisins at concentrations above the EC recommended level.

Compared to all other toxins, the worldwide contamination by ochratoxins was relatively low. Nonetheless, 2007 was a peak year for OTA contamination in which 7.2 % of all samples had a concentration above the EC recommendation of 50 ppb OTA in complementary and complete feedstuffs for pigs.

Tab. 7.2 illustrates the results of the mycotoxin survey for all European samples. Although most samples originate from central Europe and fewer from southern Europe, some differences can be observed. A more detailed overview of the results for samples originating from central and southern Europe can be found in Tab. 7.3 and Tab. 7.4, respectively.

Tab. 7.2: Results of the BIOMIN Mycotoxin Survey for all European samples.

Mycotoxin	Year	# samples	100–299 ppb	300–899 ppb	900–999 ppb	1000–1999 ppb	2000–4999 ppb	5000–7999 ppb	≥ 8000 ppb	≥ 900 ppb DON (EC recomm. pigs)
DON	2005	703	4.1 %	22.0 %	1.1 %	7.7 %	3.0 %	0.3 %	0.0 %	12.1 %
	2006	616	4.7 %	24.8 %	2.9 %	12.7 %	6.2 %	1.5 %	0.0 %	23.2 %
	2007	**484**	**11.2 %**	**29.8 %**	**2.9 %**	**11.2 %**	**6.2 %**	**1.9 %**	**3.3 %**	**25.4 %**
	2008	655	9.9 %	35.9 %	2.7 %	11.5 %	5.5 %	1.1 %	0.2 %	20.9 %
	2009	1137	11.0 %	25.9 %	1.3 %	6.1 %	2.6 %	0.1 %	0.0 %	10.1 %
	2010	1103	9.2 %	29.3 %	2.3 %	9.0 %	5.6 %	2.0 %	1.4 %	20.2 %
	2011	1481	10.3 %	30.4 %	2.1 %	9.7 %	7.1 %	0.7 %	0.5 %	20.1 %
	2012	1580	22.2 %	21.9 %	2.1 %	6.5 %	4.7 %	0.5 %	0.4 %	14.2 %
	2013	1854	17.7 %	24.1 %	1.5 %	7.8 %	5.6 %	0.3 %	0.4 %	15.5 %
	2014	736	21.7 %	27.6 %	2.4 %	7.2 %	3.3 %	0.8 %	0.1 %	13.9 %

Mycotoxin	Year	# samples	20–49 ppb	50–99 ppb	100–199 ppb	200–249 ppb	250–499 ppb	500–1999 ppb	≥ 2000 ppb	≥ 100 ppb ZEN (EC recomm. pigs)
ZEN	2004	566	2.5 %	1.8 %	1.6 %	0.5 %	1.2 %	0.5 %	0.2 %	4.1 %
	2005	378	2.4 %	8.7 %	4.2 %	1.9 %	2.9 %	0.8 %	0.0 %	9.8 %
	2006	**412**	**2.4 %**	**5.8 %**	**8.0 %**	**2.9 %**	**3.6 %**	**2.7 %**	**2.4 %**	**19.7 %**
	2007	300	14.0 %	11.0 %	6.0 %	1.3 %	3.0 %	1.7 %	0.7 %	12.7 %
	2008	413	7.3 %	9.2 %	6.1 %	0.7 %	1.2 %	0.7 %	0.0 %	8.7 %
	2009	967	5.7 %	5.6 %	4.3 %	1.3 %	2.3 %	1.2 %	0.0 %	9.2 %
	2010	759	5.5 %	6.9 %	6.7 %	1.4 %	0.9 %	0.5 %	0.0 %	9.6 %
	2011	1089	10.4 %	8.5 %	7.8 %	1.3 %	2.6 %	2.0 %	0.0 %	13.7 %
	2012	1199	11.4 %	3.8 %	2.4 %	0.3 %	1.3 %	0.5 %	0.0 %	4.5 %
	2013	1413	7.2 %	4.3 %	2.3 %	0.5 %	0.6 %	0.4 %	0.1 %	4.0 %
	2014	607	13.8 %	7.6 %	3.1 %	1.0 %	0.8 %	0.7 %	0.0 %	5.6 %

Mycotoxin	Year	# samples	1–4.9 ppb	5–9.9 ppb	10–19.9 ppb	20–199 ppb	≥ 200 ppb	≥ 5 ppb AfB_1 (EC max for dairy cattle)
Total aflatoxin B_1, B_2, G_1, G_2	2007	57	14.0 %	0.0 %	0.0 %	3.5 %	0.0 %	3.5 %
	2008	57	12.3 %	1.8 %	0.0 %	7.0 %	1.8 %	10.5 %
	2009	334	23.4 %	6.0 %	3.6 %	0.3 %	0.3 %	10.2 %
	2010	74	18.9 %	2.7 %	1.4 %	1.4 %	0.0 %	5.4 %
	2011	199	27.6 %	4.0 %	1.5 %	0.5 %	0.0 %	6.0 %
	2012	369	25.5 %	7.6 %	1.9 %	2.2 %	0.0 %	11.7 %
	2013	**711**	**20.7 %**	**7.7 %**	**3.2 %**	**3.2 %**	**0.0 %**	**14.2 %**
	2014	279	23.7 %	5.0 %	0.4 %	1.1 %	0.4 %	6.8 %

Tab. 7.2 (continued)

Mycotoxin	Year	# samples	200–999	1000–1499	1500–2999	3000–4999	5000–19 999	≥ 20 000	≥ 5000 ppb total fumonisins (EC recomm. pigs/horses)
			ppb						
Total fumonisin B_1, B_2	**2009**	**253**	**10.3 %**	**6.3 %**	**17.0 %**	**17.0 %**	**27.3 %**	**1.6 %**	**28.9 %**
	2010	70	11.4 %	2.9 %	11.4 %	7.1 %	8.6 %	0.0 %	8.6 %
	2011	229	21.8 %	3.1 %	4.4 %	5.7 %	1.3 %	0.0 %	1.3 %
	2012	358	39.1 %	0.8 %	3.6 %	1.1 %	1.1 %	0.0 %	1.1 %
	2013	584	25.0 %	2.9 %	6.8 %	3.1 %	1.5 %	0.2 %	1.7 %
	2014	314	23.6 %	3.8 %	4.5 %	2.9 %	3.2 %	0.0 %	3.2 %

Mycotoxin	Year	# samples	5–9.9	10–49	50–79	80–99	100–149	150–399	≥ 400	≥ 50 ppb OTA (EC recomm. pigs)
			ppb							
OTA	2004	25	4.0 %	4.0 %	0.0 %	0.0 %	0.0 %	0.0 %	0.0 %	0.0 %
	2005	**27**	**3.7 %**	**11.1 %**	**0.0 %**	**0.0 %**	**0.0 %**	**7.4 %**	**3.7 %**	**11.1 %**
	2006	49	0.0 %	2.0 %	2.0 %	0.0 %	0.0 %	2.0 %	0.0 %	4.1 %
	2007	12	8.3 %	0.0 %	0.0 %	0.0 %	0.0 %	0.0 %	0.0 %	0.0 %
	2008	38	0.0 %	5.3 %	2.6 %	0.0 %	0.0 %	0.0 %	0.0 %	2.6 %
	2009	139	6.5 %	2.9 %	0.0 %	0.0 %	0.0 %	2.9 %	0.0 %	2.9 %
	2010	90	3.3 %	3.3 %	0.0 %	0.0 %	0.0 %	1.1 %	0.0 %	1.1 %
	2011	261	8.0 %	6.1 %	0.0 %	0.0 %	0.0 %	0.0 %	0.0 %	0.0 %
	2012	400	9.0 %	5.0 %	1.5 %	0.0 %	0.0 %	0.0 %	0.0 %	1.5 %
	2013	614	3.3 %	6.0 %	0.3 %	0.0 %	0.5 %	0.8 %	0.2 %	1.8 %
	2014	259	2.7 %	0.4 %	0.4 %	0.0 %	0.0 %	0.0 %	0.0 %	0.4 %

Tab. 7.3: Results of the BIOMIN Mycotoxin Survey for samples sourced in central Europe.

Mycotoxin	Year	# samples	100–299	300–899	900–999	1000–1999	2000–4999	5000–7999	≥ 8000	≥ 900 ppb DON (EC recomm. pigs)
			ppb							
DON	2004	502	24.5 %	15.5 %	0.6 %	3.2 %	1.2 %	0.4 %	0.2 %	5.6 %
	2005	496	4.8 %	22.8 %	1.0 %	6.7 %	2.0 %	0.0 %	0.0 %	9.7 %
	2006	595	4.5 %	24.7 %	3.0 %	13.1 %	6.2 %	1.5 %	0.0 %	23.9 %
	2007	**478**	**11.3 %**	**29.7 %**	**2.9 %**	**11.3 %**	**6.1 %**	**1.9 %**	**3.3 %**	**25.5 %**
	2008	620	10.0 %	35.3 %	2.7 %	11.6 %	5.8 %	1.1 %	0.2 %	21.5 %
	2009	964	10.3 %	23.8 %	1.1 %	6.2 %	2.4 %	0.1 %	0.0 %	9.9 %
	2010	972	9.1 %	29.2 %	2.4 %	9.6 %	5.9 %	2.3 %	1.4 %	21.5 %
	2011	1040	10.0 %	31.3 %	1.9 %	10.8 %	7.9 %	0.8 %	0.6 %	21.9 %
	2012	1158	22.5 %	23.7 %	2.2 %	6.9 %	4.6 %	0.5 %	0.4 %	14.6 %
	2013	1331	17.3 %	26.0 %	1.5 %	8.3 %	6.5 %	0.3 %	0.1 %	16.8 %
	2014	568	23.8 %	30.1 %	3.0 %	6.2 %	2.8 %	0.7 %	0.2 %	12.9 %

Tab. 7.3 (continued)

Mycotoxin	Year	# samples	20–49	50–99	100–199	200–249	250–499	500–1999	≥ 2000	≥ 100 ppb ZEN (EC recomm. pigs)
			ppb							
ZEN	2004	480	0.8 %	0.8 %	0.8 %	0.0 %	0.6 %	0.2 %	0.2 %	1.9 %
	2005	220	4.1 %	9.1 %	5.0 %	1.4 %	1.4 %	0.5 %	0.0 %	8.2 %
	2006	**393**	**2.5 %**	**6.1 %**	**8.1 %**	**3.1 %**	**3.8 %**	**2.8 %**	**2.5 %**	**20.4 %**
	2007	295	14.2 %	11.2 %	5.8 %	1.0 %	3.1 %	1.7 %	0.7 %	12.2 %
	2008	382	7.3 %	9.4 %	6.3 %	0.8 %	1.3 %	0.8 %	0.0 %	9.2 %
	2009	814	6.0 %	6.1 %	4.8 %	1.6 %	2.7 %	1.5 %	0.0 %	10.6 %
	2010	643	6.1 %	7.3 %	7.6 %	1.6 %	1.1 %	0.6 %	0.0 %	10.9 %
	2011	669	10.9 %	10.9 %	10.0 %	1.3 %	2.7 %	1.3 %	0.0 %	15.4 %
	2012	829	12.9 %	4.8 %	2.5 %	0.2 %	1.4 %	0.4 %	0.0 %	4.6 %
	2013	930	6.8 %	4.0 %	2.4 %	0.5 %	0.4 %	0.2 %	0.2 %	3.8 %
	2014	442	12.2 %	5.2 %	2.7 %	0.7 %	1.1 %	0.2 %	0.0 %	4.8 %

Mycotoxin	Year	# samples	1–4.9	5–9.9	10–19.9	20–199	≥ 200	≥ 5 ppb AfB_1 (EC max for dairy cattle)
			ppb					
Total aflatoxin B_1, B_2, G_1, G_2	2007	52	13.5 %	0.0 %	0.0 %	0.0 %	0.0 %	0.0 %
	2008	**34**	**5.9 %**	**0.0 %**	**0.0 %**	**8.8 %**	**2.9 %**	**11.8 %**
	2009	286	25.5 %	6.3 %	4.2 %	0.3 %	0.3 %	11.2 %
	2010	29	13.8 %	0.0 %	0.0 %	0.0 %	0.0 %	0.0 %
	2011	26	15.4 %	0.0 %	0.0 %	0.0 %	0.0 %	0.0 %
	2012	119	10.1 %	1.7 %	2.5 %	2.5 %	0.0 %	6.7 %
	2013	357	15.7 %	3.6 %	3.9 %	3.6 %	0.0 %	11.2 %
	2014	169	21.3 %	0.0 %	0.0 %	0.6 %	0.0 %	0.6 %

Mycotoxin	Year	# samples	200–999	1000–1499	1500–2999	3000–4999	5000–19 999	≥ 20 000	≥ 5000 ppb total fumonisins (EC recomm. pigs/horses)
			ppb						
Total fumonisin B_1, B_2	2007	42	45.2 %	11.9 %	19.0 %	11.9 %	4.8 %	0.0 %	4.8 %
	2008	15	13.3 %	0.0 %	13.3 %	0.0 %	26.7 %	0.0 %	26.7 %
	2009	**233**	**11.2 %**	**5.6 %**	**16.3 %**	**17.2 %**	**28.8 %**	**1.7 %**	**30.5 %**
	2010	41	9.8 %	0.0 %	0.0 %	0.0 %	9.8 %	0.0 %	9.8 %
	2011	88	19.3 %	4.5 %	4.5 %	0.0 %	0.0 %	0.0 %	0.0 %
	2012	111	4.5 %	0.0 %	0.0 %	0.0 %	0.0 %	0.0 %	0.0 %
	2013	290	10.3 %	2.4 %	2.1 %	0.7 %	0.0 %	0.0 %	0.0 %
	2014	200	19.0 %	2.0 %	1.0 %	0.0 %	1.0 %	0.0 %	1.0 %

Tab. 7.3 (continued)

Mycotoxin	Year	# samples	5–9.9	10–49	50–79	80–99	100–149	150–399	≥ 400	≥ 50 ppb OTA (EC recomm. pigs)
			ppb							
OTA	2004	14	0.0 %	0.0 %	0.0 %	0.0 %	0.0 %	0.0 %	0.0 %	0.0 %
	2005	**14**	**7.1 %**	**7.1 %**	**0.0 %**	**0.0 %**	**0.0 %**	**14.3 %**	**0.0 %**	**14.3 %**
	2006	27	0.0 %	0.0 %	0.0 %	0.0 %	0.0 %	0.0 %	0.0 %	0.0 %
	2007	9	0.0 %	0.0 %	0.0 %	0.0 %	0.0 %	0.0 %	0.0 %	0.0 %
	2008	24	0.0 %	0.0 %	0.0 %	0.0 %	0.0 %	0.0 %	0.0 %	0.0 %
	2009	119	6.7 %	2.5 %	0.0 %	0.0 %	0.0 %	3.4 %	0.0 %	3.4 %
	2010	46	0.0 %	6.5 %	0.0 %	0.0 %	0.0 %	2.2 %	0.0 %	2.2 %
	2011	76	3.9 %	2.6 %	0.0 %	0.0 %	0.0 %	0.0 %	0.0 %	0.0 %
	2012	135	3.7 %	3.0 %	2.2 %	0.0 %	0.0 %	0.0 %	0.0 %	2.2 %
	2013	313	1.9 %	8.0 %	0.0 %	0.0 %	0.0 %	0.3 %	0.0 %	0.3 %
	2014	162	1.2 %	0.0 %	0.6 %	0.0 %	0.0 %	0.0 %	0.0 %	0.6 %

Tab. 7.4: Results of the BIOMIN Mycotoxin Survey for samples sourced in southern Europe.

Mycotoxin	Year	# samples	100–299	300–899	900–999	1000–1999	2000–4999	5000–7999	≥ 8000	≥ 900 ppb DON (EC recomm. pigs)
			ppb							
DON	2008	33	9.1 %	45.5 %	3.0 %	9.1 %	0.0 %	0.0 %	0.0 %	12.1 %
	2009	281	9.6 %	12.5 %	2.1 %	1.4 %	1.1 %	0.0 %	0.0 %	4.6 %
	2010	63	14.3 %	20.6 %	0.0 %	4.8 %	1.6 %	0.0 %	0.0 %	6.3 %
	2011	147	8.2 %	17.7 %	1.4 %	4.1 %	4.8 %	0.0 %	0.0 %	10.2 %
	2012	215	19.5 %	7.0 %	0.5 %	1.9 %	0.5 %	0.0 %	0.5 %	3.3 %
	2013	342	21.3 %	12.6 %	0.0 %	3.8 %	0.6 %	0.0 %	0.3 %	4.7 %
	2014	**84**	**7.1 %**	**20.2 %**	**1.2 %**	**9.5 %**	**3.6 %**	**0.0 %**	**0.0 %**	**14.3 %**
Mycotoxin	**Year**	**# samples**	**20–49**	**50–99**	**100–199**	**200–249**	**250–499**	**500–1999**	**≥ 2000**	**≥ 100 ppb ZEN (EC recomm. pigs)**
			ppb							
ZEN	**2009**	**268**	**7.1 %**	**9.0 %**	**11.6 %**	**4.5 %**	**7.1 %**	**2.2 %**	**0.0 %**	**25.4 %**
	2010	47	4.3 %	8.5 %	4.3 %	2.1 %	0.0 %	0.0 %	0.0 %	6.4 %
	2011	133	5.3 %	3.0 %	2.3 %	0.0 %	2.3 %	1.5 %	0.0 %	6.0 %
	2012	193	4.1 %	0.5 %	3.1 %	0.0 %	0.5 %	1.0 %	0.0 %	4.7 %
	2013	306	6.2 %	3.3 %	2.3 %	0.3 %	0.7 %	0.3 %	0.0 %	3.6 %
	2014	80	17.5 %	6.3 %	3.8 %	2.5 %	0.0 %	3.8 %	0.0 %	10.0 %

Tab. 7.4 (continued)

Mycotoxin	Year	# samples	1–4.9 ppb	5–9.9 ppb	10–19.9 ppb	20–199 ppb	≥ 200 ppb	≥ 5 ppb AfB_1 (EC max for dairy cattle)
Total aflatoxin B_1, B_2, G_1, G_2	2008	21	23.8 %	4.8 %	0.0 %	4.8 %	0.0 %	9.5 %
	2009	253	28.9 %	7.9 %	4.7 %	0.4 %	0.4 %	13.4 %
	2010	34	29.4 %	5.9 %	2.9 %	2.9 %	0.0 %	11.8 %
	2011	126	24.6 %	3.2 %	2.4 %	0.8 %	0.0 %	6.3 %
	2012	173	29.5 %	3.5 %	1.7 %	2.9 %	0.0 %	8.1 %
	2013	**286**	**29.7 %**	**11.9 %**	**3.1 %**	**2.8 %**	**0.0 %**	**17.8 %**
	2014	55	40.0 %	1.8 %	1.8 %	3.6 %	1.8 %	9.1 %

Mycotoxin	Year	# samples	200–999 ppb	1000–1499 ppb	1500–2999 ppb	3000–4999 ppb	5000–19 999 ppb	≥ 20 000 ppb	≥ 5000 ppb total fumonisins (EC recomm. pigs/horses)
Total fumonisin B_1, B_2	**2009**	**225**	**10.2 %**	**7.1 %**	**18.2 %**	**19.1 %**	**30.2 %**	**1.8 %**	**32.0 %**
	2010	24	8.3 %	8.3 %	33.3 %	20.8 %	8.3 %	0.0 %	8.3 %
	2011	105	25.7 %	2.9 %	4.8 %	10.5 %	1.9 %	0.0 %	1.9 %
	2012	166	60.2 %	0.6 %	7.2 %	2.4 %	2.4 %	0.0 %	2.4 %
	2013	231	39.8 %	3.9 %	11.7 %	6.1 %	3.9 %	0.4 %	4.3 %
	2014	65	33.8 %	7.7 %	18.5 %	12.3 %	10.8 %	0.0 %	10.8 %

It is noteworthy that more than 10 % of samples tested each year originating from different European countries, especially from central Europe, showed levels of DON exceeding the EC recommended level of 900 ppb. In 2006, 2008, 2010, and 2011 more than 20 % of the samples tested exceeded this level, and 2007 can be considered the DON-contamination peak year, as one-fourth of the samples was above 900 ppb. Compared to worldwide results, the contamination of ZEN in all European samples, mainly originating from central Europe, was relatively lower. However, between 4.1 % and 19.7 % of all samples was above the EC recommended level of 100 ppb. Comparable with the global results, the ZEN-contamination peak year in Europe was also 2006.

Until a few years ago, aflatoxins were not considered an actual threat in Europe. However, the results from the mycotoxin survey illustrate that 3.5 % to 14.2 % of European samples contain concentrations above the European guideline of 5 ppb aflatoxin B_1 for feed intended for dairy. As expected, Europe experienced the highest aflatoxin contamination in 2013. In the first quarter of 2013, several European countries, including Romania, Serbia, and Croatia, reported that milk for human consumption had been contaminated with aflatoxins. Thereafter, feed originating from Serbia imported to the Netherlands and Germany appeared to be contaminated with aflatoxins. Fumonisins are also not recognized as a threat in Europe. Nonetheless, in 2009 almost 29 % of the European samples tested contained concentrations above the 5000 ppb EC recommended level. The contamination of ochratoxins in Europe showed a peak in

2005 with 11.1 % of samples containing concentrations above the EC recommendation of 50 ppb for feed intended for pigs.

The results for samples sourced in central Europe presented in Tab. 7.3 are very similar to those for all European samples. The results show that 2008, 2009, and 2013 were peak years in terms of aflatoxin contamination, as the concentration of more than 11 % of the samples tested in these periods were above the 5 ppb EU guidance level for feed for dairy cows. These results should raise the awareness of the problem of aflatoxins in a region where these toxins were not previously considered a problem.

As expected, the aflatoxin contamination was higher in southern Europe than in central European countries. The highest concentrations were observed in 2013, as the concentration of almost 18 % of all samples tested was above the 5 ppb EU maximum level (Tab. 7.4). The fumonisin levels in southern European countries were also higher than in central Europe. In 2009, the concentrations of zearalenone were at a peak in southern Europe, with 25 % of samples above the 100 ppb EC recommendation for feed intended for pigs. The results for the first half of 2014 show high concentrations of ZEN and DON in this region. The co-occurrence of zearalenone and DON is a known phenomenon, as both mycotoxins are produced by the same fungus.

In the past decade, a total of 2738 corn samples (Tab. 7.5) and 1714 wheat samples (Tab. 7.6) originating from Europe have been analyzed for their mycotoxin content. European wheat showed the highest DON contamination in 2005, 2008, 2012, and 2013. During these years more than 20 % of the samples contained DON at concentrations above the EC recommended value of 900 ppb. The DON-contamination peak was observed in the first half of 2014, when the concentrations of almost 36 % of the samples tested were above 900 ppb. The highest aflatoxin concentrations in wheat were observed in 2009 with almost 15 % of the samples above the EU maximum of 5 ppb for feed intended for dairy cows.

Tab. 7.5: Results of the BIOMIN Mycotoxin Survey for corn and corn silage samples sourced in Europe.

Mycotoxin	Year	# samples	100–299	300–899	900–999	1000–1999	2000–4999	5000–7999	≥ 8000	≥ 900 ppb DON (EC recomm. pigs)
			ppb							
	2004	148	25.0 %	22.3 %	1.4 %	4.1 %	0.0 %	0.7 %	0.0 %	6.1 %
	2005	192	7.3 %	27.6 %	1.0 %	10.9 %	6.8 %	0.0 %	0.0 %	18.8 %
	2006	**214**	**1.4 %**	**36.0 %**	**6.5 %**	**21.5 %**	**11.7 %**	**2.3 %**	**0.0 %**	**42.1 %**
	2007	159	6.3 %	44.7 %	5.0 %	18.2 %	9.4 %	0.6 %	1.9 %	35.2 %
	2008	234	9.4 %	37.6 %	2.1 %	16.2 %	9.4 %	2.1 %	0.4 %	30.3 %
DON	2009	280	13.2 %	36.1 %	1.1 %	6.8 %	4.3 %	0.4 %	0.0 %	12.5 %
	2010	254	7.9 %	31.5 %	2.0 %	14.2 %	11.0 %	8.7 %	2.8 %	38.6 %
	2011	400	7.5 %	31.3 %	2.0 %	15.0 %	14.0 %	1.8 %	0.8 %	33.5 %
	2012	329	24.0 %	25.8 %	3.6 %	5.8 %	2.1 %	0.0 %	0.3 %	11.9 %
	2013	405	12.1 %	26.2 %	0.7 %	10.4 %	6.4 %	0.7 %	0.0 %	18.3 %
	2014	193	10.4 %	27.5 %	4.7 %	7.3 %	2.6 %	0.5 %	0.0 %	15.0 %

Tab. 7.5 (continued)

Mycotoxin	Year	# samples	20–49 ppb	50–99 ppb	100–199 ppb	200–249 ppb	250–499 ppb	500–1999 ppb	≥ 2000 ppb	≥ 100 ppb ZEN (EC recomm. pigs)
ZEN	2004	139	1.4 %	1.4 %	1.4 %	0.0 %	1.4 %	0.0 %	0.0 %	2.9 %
	2005	70	2.9 %	12.9 %	10.0 %	2.9 %	7.1 %	2.9 %	0.0 %	22.9 %
	2006	**119**	**4.2 %**	**16.0 %**	**19.3 %**	**6.7 %**	**9.2 %**	**2.5 %**	**1.7 %**	**39.5 %**
	2007	85	11.8 %	18.8 %	12.9 %	2.4 %	5.9 %	2.4 %	0.0 %	23.5 %
	2008	125	11.2 %	8.0 %	11.2 %	2.4 %	2.4 %	1.6 %	0.0 %	17.6 %
	2009	200	2.5 %	7.0 %	3.5 %	0.5 %	0.5 %	0.0 %	0.0 %	4.5 %
	2010	188	9.0 %	13.3 %	19.1 %	3.7 %	2.1 %	1.1 %	0.0 %	26.1 %
	2011	241	12.9 %	7.5 %	12.0 %	1.2 %	3.7 %	3.7 %	0.0 %	20.7 %
	2012	248	8.1 %	4.4 %	1.6 %	0.4 %	1.6 %	0.8 %	0.0 %	4.4 %
	2013	338	12.1 %	8.3 %	5.9 %	1.5 %	2.4 %	1.2 %	0.6 %	11.5 %
	2014	164	11.6 %	14.0 %	4.3 %	1.8 %	1.8 %	0.6 %	0.0 %	8.5 %

Mycotoxin	Year	# samples	1–4.9 ppb	5–9.9 ppb	10–19.9 ppb	20–199 ppb	≥ 200 ppb	≥ 5 ppb AfB_1 (EC max for dairy cattle)
Total aflatoxin B_1, B_2, G_1, G_2	2011	51	29.4 %	2.0 %	3.9 %	2.0 %	0.0 %	7.9 %
	2012	74	17.6 %	4.1 %	2.7 %	1.4 %	0.0 %	8.1 %
	2013	**213**	**11.3 %**	**5.2 %**	**4.7 %**	**3.3 %**	**0.0 %**	**13.1 %**
	2014	83	28.9 %	0.0 %	0.0 %	3.6 %	0.0 %	3.6 %

Mycotoxin	Year	# samples	200–999 ppb	1000–1499 ppb	1500–2999 ppb	3000–4999 ppb	5000–19 999 ppb	≥ 20 000 ppb	≥ 5000 ppb total fumonisins (EC recomm. pigs/horses)
Total fumonisin B_1, B_2	**2011**	**72**	**31.9 %**	**8.3 %**	**8.3 %**	**8.3 %**	**11.1 %**	**0.0 %**	**11.1 %**
	2012	63	36.5 %	0.0 %	12.7 %	4.8 %	3.2 %	0.0 %	3.2 %
	2013	169	21.3 %	4.1 %	13.6 %	7.7 %	3.6 %	0.6 %	4.1 %
	2014	82	24.4 %	7.3 %	6.1 %	7.3 %	1.2 %	0.0 %	1.2 %

Mycotoxin	Year	# samples	5–9.9 ppb	10–49 ppb	50–79 ppb	80–99 ppb	100–149 ppb	150–399 ppb	≥ 400 ppb	≥ 50 ppb OTA (EC recomm. pigs)
OTA	2011	41	2.4 %	9.8 %	0.0 %	0.0 %	0.0 %	0.0 %	0.0 %	0.0 %
	2012	**85**	**1.2 %**	**3.5 %**	**1.2 %**	**0.0 %**	**0.0 %**	**0.0 %**	**0.0 %**	**1.2 %**
	2013	168	3.0 %	3.0 %	0.0 %	0.0 %	0.0 %	0.0 %	0.0 %	0.0 %
	2014	79	0.0 %	0.0 %	0.0 %	0.0 %	0.0 %	0.0 %	0.0 %	0.0 %

Tab. 7.6: Results of the BIOMIN Mycotoxin Survey for wheat samples sourced in Europe.

Mycotoxin	Year	# samples	100–299	300–899	900–999	1000–1999	2000–4999	5000–7999	≥ 8000	≥ 900 ppb DON (EC recomm. pigs)
			ppb							
DON	2004	98	34.7 %	28.6 %	1.0 %	4.1 %	7.1 %	1.0 %	1.0 %	14.3 %
	2005	87	5.7 %	21.8 %	3.4 %	16.1 %	2.3 %	0.0 %	0.0 %	21.8 %
	2006	95	10.5 %	16.8 %	0.0 %	9.5 %	2.1 %	0.0 %	0.0 %	11.6 %
	2007	66	1.5 %	19.7 %	3.0 %	6.1 %	1.5 %	4.5 %	1.5 %	16.7 %
	2008	159	3.8 %	41.5 %	5.0 %	10.7 %	6.9 %	1.3 %	0.0 %	23.9 %
	2009	374	5.6 %	18.4 %	1.1 %	4.5 %	2.1 %	0.0 %	0.0 %	7.8 %
	2010	194	6.7 %	28.4 %	3.1 %	7.2 %	4.6 %	0.0 %	2.1 %	17.0 %
	2011	202	5.4 %	31.7 %	2.0 %	5.9 %	3.0 %	0.5 %	0.5 %	11.9 %
	2012	253	13.8 %	32.4 %	3.2 %	11.1 %	10.3 %	2.0 %	0.8 %	27.3 %
	2013	332	13.0 %	28.0 %	1.5 %	12.3 %	8.7 %	0.3 %	0.3 %	23.2 %
	2014	**56**	**14.3 %**	**19.6 %**	**3.6 %**	**17.9 %**	**10.7 %**	**3.6 %**	**0.0 %**	**35.7 %**

Mycotoxin	Year	# samples	20–49	50–99	100–199	200–249	250–499	500–1999	≥ 2000	≥ 100 ppb ZEN (EC recomm. pigs)
			ppb							
ZEN	2004	96	6.3 %	5.2 %	2.1 %	2.1 %	1.0 %	1.0 %	1.0 %	7.3 %
	2005	54	3.7 %	13.0 %	0.0 %	5.6 %	7.4 %	1.9 %	0.0 %	14.8 %
	2006	63	3.2 %	0.0 %	4.8 %	0.0 %	0.0 %	0.0 %	0.0 %	4.8 %
	2007	37	2.7 %	0.0 %	0.0 %	0.0 %	2.7 %	0.0 %	0.0 %	2.7 %
	2008	103	5.8 %	17.5 %	4.9 %	0.0 %	1.9 %	1.0 %	0.0 %	7.8 %
	2009	**346**	**5.5 %**	**6.9 %**	**8.4 %**	**3.5 %**	**5.5 %**	**1.7 %**	**0.0 %**	**19.1 %**
	2010	97	3.1 %	3.1 %	4.1 %	0.0 %	1.0 %	0.0 %	0.0 %	5.2 %
	2011	144	2.1 %	6.3 %	6.9 %	1.4 %	1.4 %	0.7 %	0.0 %	10.4 %
	2012	162	12.3 %	3.1 %	6.2 %	0.6 %	2.5 %	1.2 %	0.0 %	10.5 %
	2013	213	1.9 %	1.4 %	0.5 %	0.0 %	0.0 %	0.0 %	0.0 %	0.5 %
	2014	45	4.4 %	2.2 %	6.7 %	0.0 %	0.0 %	0.0 %	0.0 %	6.7 %

Mycotoxin	Year	# samples	1–4.9	5–9.9	10–19.9	20–199	≥ 200	≥ 5 ppb AfB_1 (EC max for dairy cattle)
			ppb					
Total aflatoxin B_1, B_2, G_1, G_2	**2009**	**222**	**32.0 %**	**8.6 %**	**5.4 %**	**0.5 %**	**0.5 %**	**14.9 %**
	2010	14	14.3 %	0.0 %	0.0 %	0.0 %	0.0 %	0.0 %
	2011	22	36.4 %	0.0 %	0.0 %	0.0 %	0.0 %	0.0 %
	2012	15	13.3 %	6.7 %	0.0 %	0.0 %	0.0 %	6.7 %
	2013	95	12.6 %	0.0 %	0.0 %	0.0 %	0.0 %	0.0 %
	2014	24	16.7 %	0.0 %	0.0 %	0.0 %	0.0 %	0.0 %

Tab. 7.6 (continued)

Mycotoxin	Year	# samples	5–9.9	10–49	50–79	80–99	100–149	150–399	≥ 400	> 50 ppb (EC recomm. Pigs)
			ppb							
OTA	**2010**	**19**	**5.3 %**	**0.0 %**	**0.0 %**	**0.0 %**	**0.0 %**	**5.3 %**	**0.0 %**	**5.3 %**
	2011	23	0.0 %	0.0 %	0.0 %	0.0 %	0.0 %	0.0 %	0.0 %	0.0 %
	2012	30	6.7 %	3.3 %	0.0 %	0.0 %	0.0 %	0.0 %	0.0 %	0.0 %
	2013	94	1.1 %	0.0 %	0.0 %	0.0 %	0.0 %	0.0 %	0.0 %	0.0 %
	2014	24	0.0 %	0.0 %	0.0 %	0.0 %	0.0 %	0.0 %	0.0 %	0.0 %

In European corn and corn silage, the results for DON and ZEN showed similar trends due to their common co-occurrence. In 2006, 2007, 2008, 2010, and 2011, over 30 % of these samples contained concentrations of DON above the EC recommendation of 900 ppb. Accordingly, ZEN contamination was also high in the years mentioned. The highest concentrations of both DON and ZEN were detected in 2006.

7.7 Conclusion

Food safety is undoubtedly linked to a complex web of different factors and would be incomplete without taking the effects caused by climate change into consideration. Weather extremes are clearly affecting fungal profiles, and consequently mycotoxin patterns, on a worldwide scale and global trade makes mycotoxin prediction even more difficult than it currently is. An integrated systems approach would be necessary to warrant the accuracy of predictions. Constant monitoring and continual research on the prevention and mitigation of mycotoxin contamination are therefore necessary. The first steps towards preventing the negative effects of these harmful substances are the implementation of good agricultural practices and proper storage conditions.

References

[1] EFSA 2012. *Modelling, Predicting and Mapping the Emergence of Aflatoxins in Cereals in the EU Due to Climate Change*. EFSA Scientific Report; 2012.

[2] Miraglia M, Marvin HJP, Kleter GA. Climate change and food safety: An emerging issue with special focus on Europe. *Food and Chemical Toxicology* 2009;47:1009–1021.

[3] Magan N, Medina A, Aldred D. Possible climate-change effects on mycotoxin contamination of food crops pre- and postharvest. *Plant Pathology* 2011;60:150–163.

[4] IPCC 2014. *Climate Change 2014: Impacts, Adaptation, and Vulnerability*. IPCC Report 2014.

[5] Rosenzweig C, Yang XB, Anderson P, Epstein P, Vicarelli M. Agriculture: climate change, crop pests and diseases. In: Epstein P,Mills E, eds *Climate Change Futures: Health, Ecological and Economic Dimensions* 2005:70–77.

[6] Chakraborty S, Newton AC. Climate change, plant diseases and food security: an overview. *Plant Pathology* 2011;60:2–14.
[7] Bebber D, Ramotowski MAT, Gurr SJ. Crop pests and pathogens move polewards in a warming world. *Nature climate change* 2013.
[8] Burrows M, Schoeman DS, Buckley LB. The pace of shifting climate in marine and terrestrial ecosystems. *Science* 2011;334:652–655.
[9] FA2008. *Climate change: Implications for food safety*. FAO Report 2008.
[10] Miller D, Richardson SN. *Mycotoxins in Canada: A perspective for 2013*. Regulatory Governance Initiative Report 2013.
[11] Magan N. Fungi in extreme environments. In: Kubicek CP, Druzhinina IS, eds *Environmental and Microbial Relationships TheMYCOTA IV* 2nd ed. Berlin, Germany: Springer Verlag 2007:85–103.
[12] Sanchis V, Magan N. Environmental profiles for growth and mycotoxin production. In: Magan N, Olsen M (eds.). *Mycotoxins in Food: Detection and Control* Cambridge, UK: Woodhead Publishing Ltd 2004:174–189.
[13] Sant'Ana AS. Special issue on climate change and food science. *Food Research International* 2010;43:1727–1728.
[14] Paterson R, Lima N. Further mycotoxin effects from climate change. *Food Research International* 2011;44:2555–2566.
[15] Garrett K, Dendy SP, Frank EE, Rouie MN, Travers SE. Climate change effects on plant disease: genomes to ecosystems. *Annual Review of Phytopathology* 2006;44:489–509.
[16] Boken V, Hoogenboom G, Williams JH, Diarra B, Dione S, Easson GL. Monitoring peanut contamination in Mali (Africa) using the AVHRR satellite data and a crop simulation model. *International Journal of Remote Sensing* 2008;29:117–129.
[17] Chauhan Y, Wright GC, Rachaputi NC. Modelling climatic risks of aflatoxin contamination in maize. *Australian Journal of Experimental Agriculture* 2008;48:358–366.
[18] Paterson R, Lima N. How will climate change affect mycotoxins in food? *Food Research International* 2010;43:1902–1914.
[19] Schaafsma A, Hooker DC. Climatic models to predict occurrence of *Fusarium* toxins in wheat and maize. *International Journal of Food Microbiology* 2007;119:116–125.
[20] Giorni P, Battilani P, Magan N. Effect of solute and matric potential on in vitro growth and sporulation of strains from a new population of *Aspergillus flavus* isolated in Italy. *Fungal Ecology* 2008;1:102–106.
[21] Ward T, Clear RM, Rooney AP. An adaptive evolutionary shift in fusarium head blight pathogen populations is driving the rapid spread of more toxigenic *Fusarium graminearum* in North America. *Fungal Genetics and Biology* 2008;45:473–484.
[22] Battilani P, Barbano C, Marin S, Sanchis V, Kozakiewicz Z, Magan N. Mapping of *Aspergillus* section Nigri in Southern Europe and Israel based on geostatistical analysis. *International Journal of Food Microbiology* 2006;111:72–82.
[23] Rossi V, Giosue S, Delogu G. A model estimating risk for *Fusarium* mycotoxins in wheat kernels. *Aspects of Applied Biology* 2003a;68:229–234.
[24] Rossi V, Giosue S, Pattori E, Spanna F, Del Vechio A. A model estimating the risk of fusarium head blight in wheat. *EPPO Bulletin* 33 2003b:421–425.
[25] de la Campa R, Hooker DC, Miller JD, Schaafsma AW, Hammond BG. Modelling the effects of environment, insect damage, and Bt genotypes on fumonisin accumulation inmaize in Argentina and the Phillipines. *Mycopathologia* 2005;159:539–552.
[26] van der Fels-Klerx H, Kandhai MC, Brynestad S. Development of a European system for identification of emerging mycotoxins in the wheat supply chain. *World Mycotoxin Journal* 2009;2:119.

[27] Beniston M, Diaz HF. The 2003 heat wave as an example of summers in a greenhouse climate? Observations and climate model simulations for Basel, Switzerland. *Global and Planetary Change* 2004;44:73–81.
[28] Tirado M, Clarke R, Jaykus LA, McQuatters-Gollop A, Frank JM. Climate change and food safety: A review. *Food Research International* 2010;43:1745–1765.
[29] Medina A, Mateo R, Valle-Algarra FM, Mateo E, Jimenez M. Effect of carbendazim and physicochemical factors on the growth and ochratoxin A production of *Aspergillus carbonarius* isolated from grapes. *International Journal of Food Microbiology* 2007a;119:230–235.
[30] Medina A, Jimenez M, Mateo R, Magan N. Natamycin efficacy for control of growth and ochratoxin production by *Aspergillus carbonarius* isolates under different environmental regimes. *Journal of Applied Microbiology* 2007b;103:2234–2239.
[31] Schmidt-Heydt M, Magan N, Geisen R. Stress induction of mycotoxin biosynthesis genes in relation to abiotic factors. *FEMS Microbiology Letters* 2008;284:142–149.
[32] Schmidt-Heydt M, Abdel-Hadi A, Magan N, Geisen R. Complex regulation of the aflatoxin biosynthesis gene cluster of *A. flavus* in relation to various combinations of water activity and temperature. *International Journal of Food Microbiology* 2009;135:231–237.
[33] Serra R, Lourenço A, Alípio P, Venâncio A. Influence of the region of origin on the mycobiota of grapes with emphasis on *Aspergillus* and *Penicillium* species. *Mycological Research* 2006;110:971–978.
[34] Serra R, Mendonca C, Venancio A. Ochratoxin A occurrence and formation in Portuguese wine grapes at various stages of maturation. *International Journal of Food Microbiology* 2006;111:35–39.
[35] Varga J, Koncz Z, Kocsube S. Mycobiota of grapes collected in Hungarian and Czech vineyards in 2004. *Acta Alimentaria* 2007;36:329–341.
[36] European Commission 2006. *COMMISSION RECOMMENDATION of 17 August 2006 on the presence of deoxynivalenol, zearalenone, ochratoxin A, T-2 and HT-2 and fumonisins in products intended for animal feeding.* 2006.
[37] European Parliament 2002. *DIRECTIVE 2002/32/EC OF THE EUROPEAN PARLIAMENT AND OF THE COUNCIL of 7 May 2002 on undesirable substances in animal feed.* 2002.

María J. Sainz, Amparo Alfonso, and Luis M. Botana

8 Considerations about international mycotoxin legislation, food security, and climate change

8.1 Introduction

Mycotoxins are low molecular-weight products (250 to 720 Da) with different chemical structure and activity. These compounds are secondary metabolites produced by species of several fungal genera, primarily *Fusarium*, *Aspergillus*, *Penicillium*, *Claviceps*, and *Alternaria*, which can colonize plants cultivated for human and/or animal consumption. The infection can occur either in the field or after harvesting, during storage, transport, and processing. Warm temperatures and high humidity are key factors for growth of mycotoxigenic fungi. This is why food and feed contaminated with mycotoxins are most frequently found in the warmer agricultural regions of the world, and also why climate change is expected to strongly affect their activity and impact on plant production and quality [77].

More than 300 mycotoxins have been identified, but about 20 can be present naturally in food and feeds at significant levels and frequently enough to become a food safety concern [76]. Seven classes of mycotoxins frequently occur: aflatoxins, ochratoxin A, patulin, trichothecenes, zearalenone, fumonisins, and ergot alkaloids. These compounds are not affected by temperature or long-term storage and are chemically stable under manufacturing conditions. Different toxic effects, acute or chronic, have been associated with these fungal metabolites, such as carcinogenicity, nephrotoxicity, hepatotoxicity, reproductive problems, gastrointestinal effects, immunosuppression, skin problems, and central nervous system disorders. Mycotoxins therefore represent a major concern for human health and cause significant economic losses in many production areas. Most countries have developed continuous monitoring programs and adopted regulations to limit exposure to these compounds in order to avoid intoxication because mycotoxins contaminate food and feed, both raw and processed materials. In order to prevent and reduce risks for consumers and to minimize the economic impact caused by the presence of these substances, public administrations pass regulations regarding permissible levels of mycotoxins in foodstuffs and official controls. To this end, a broad range of analytical techniques are available for detection and quantification of mycotoxins, however highly sensitive methods are still necessary in order to avoid food poisoning and to guarantee the detection of known and new contaminants. Besides, mycotoxins are not homogeneously distributed in product lots; in addition to the detection method, the sampling plan therefore plays a crucial part in identifying these products. Thus, the criteria that both sampling and detection methods should comply have already been fixed by several regulations in the European Union [74, 75]. In the case of detection methods, the performance criteria are independent of the method used and are addressed to ensure that control laboratories use analysis techniques with minimum performance characteristics to achieve comparable levels of operation, since no official detection methods have been established. This is a fundamental difference compared to marine toxins, basically because mycotoxins pose a chronic risk, while phycotoxins are mostly an acute food risk [73]; a specific validated method has been implemented for marine toxins in the Commission Regulation (EU) No 15/2011 of January 10, 2011 [72].

8.1.1 Main mycotoxins

Aflatoxins are produced by several species of *Aspergillus* genus, manly *A. flavus* and *A. parasiticus*. This group of compounds includes aflatoxin B_1, B_2, G_1, and G_2. Aflatoxins are difuranocoumarin derivatives which can be metabolized in the liver through the cytochrome-P450 family to epoxide or hydroxylated derivatives. These toxins are found primarily in warm and humid climates and contaminate crops and raw materials of cereals, oilseeds, spices, fruits, cottonseed, and nuts [71]. Aflatoxin B_1 is the most toxic and carcinogenic to humans and animals, with an LD_{50} (mean lethal dose; lethal dose, 50 %) of between 0.3 and 18 mg/kg depending on the route of administration and the animal species [70]. When ruminants are fed with aflatoxin B_1-contaminated feedstuffs, AFB_1 is metabolized in the liver to secondary metabolites. The major metabolite is the hydroxylated metabolite aflatoxin M_1 (AFM_1) which is excreted into milk and can be found in milk and other dairy products [69]. AFM_1 is considered a possible human carcinogen by the IARC.

Ochratoxin A is a phenylalanine-dihydroisocoumarin derivative produced by *Aspergillus* and *Penicillium* genera. This compound is found as a contaminant in crops and raw materials of cereals, spices, cocoa, coffee, dry grapes, wine, and pork and chicken meat [71]. This toxin is structurally similar to the amino acid phenylalanine and can therefore participate in cellular pathways where the amino acid is involved [68]. The LD_{50} of ochratoxin A ranges from 50 mg/kg for dogs to 0.5 mg/kg for mice [70].

Trichothecenes are tetracyclic sesquiterpenoids produced by several *Fusarium* species. There are more than 170 identified compounds divided into four types, A–D. The main representatives are HT-2 and T-2 toxins, from type A group, and deoxynivalenol from type B group. These toxins can be found in cereals and toxic effects such as gastroenteritis, intestinal hemorrhage, immunosuppression, and dermatotoxicity have been associated with exposure to them.

Fumonisins are a group of compounds produced by several *Fusarium* species found in maize from warm regions. These toxins, structurally similar to the sphingolipid precursor sphinganine, interfere with sphingosine synthesis leading to cytotoxic compounds which are carcinogenic and neurotoxic. At least 12 fumonisin analogs are known. Fumonisin B_1, B_2, and B_3 are the most important compounds within this group.

Zearalenone is a lactone produced by several *Fusarium* species, found as a contaminant in cereals and maize oil with estrogenic activity.

Patulin is an unsaturated heterocyclic lactone mainly produced by *Penicillium expansum*, but also by species of *Aspergillus*, *Byssochlamys*, *Eupenicillium*, and *Paecilomyces*. This compound has high affinity with sulfhydryl groups and consequently is involved in several enzymatic processes. Patulin induces acute symptoms such as lung and brain edema, liver, spleen and kidney damage, and is toxic for the immune system. In addition, patulin is related to protein synthesis inhibition. The LD_{50} oscillates between 15 and 25 mg/kg [70]. This toxin has been detected in apples and apple products, although high contamination levels have rarely been reported.

Species of the genus *Claviceps*, notably *C. purpurea*, infect seed heads, usually of rye, replacing the developing grain with the sclerotium of ergot which contains alkaloids.

Ergot alkaloids are tetracyclic compounds, derived from 6-methylergoline, which are produced by *Claviceps* species. These mycotoxins can be partial agonist or antagonist of adrenergic, dopaminergic, and serotoninergic receptors. Ischemias of the extremities, miscarriage, and central disorders have been described after consumption of grain infected by the fungi.

Alternaria species, particularly *A. alternata*, infect a wide range of crops, such as cereals, both pre- and postharvest, and produce alternariol, alternariol monomethyl ether, altertoxins I–III, altenuene, and tenuazonic acid. Their toxic effects are complex, but esophageal cancer has been proposed after epidemiologic studies in China [67].

In addition to trichothecenes, zearalenone, and fumonisins, other emerging *Fusarium* toxins such as fusaproliferin, enniatins, beauvericin, moniliformin, and modified forms of DON, ZON, NIV, FUM, and T-2 and HT-2 toxins have been reported. These compounds have different structures and are not yet regulated, although several toxic effects have been described after cellular exposition of these toxins.

8.2 Impacts of climate change on agriculture

The observed increases in average global air and ocean temperatures, widespread melting of snow and ice, and the rising of average global sea level since the beginning of the 20th century are evidence of the process of global warming [66]. Global land-ocean temperature in the decade 2000–2010 was about 0.8 °C warmer than at the beginning of the 20th century, and two thirds of the warming occurred since 1975 [65]. Further to this, global climate models have shown a greater than 90 % probability that the hottest seasons on record will represent the future norm in many temperate locations, and that growing season temperatures in the tropics and subtropics will exceed the most extreme seasonal temperatures recorded in the 20th century [64]. All cropland areas are thus expected to experience some degree of warming, with the largest change being predicted in high latitudes, and small increases in temperature, although probably with a significant impact on agriculture, in low latitudes [63].

Currently ca 1600 Mha of the world are cultivated, of which 1300 Mha (80 %) are rainfed and produce about 60 % of the global crop output, with the most productive systems concentrated in the temperate zones of Europe, followed by northern America and areas in the subtropics and humid tropics [62]. Irrigated agriculture account for the remaining 20 % of arable land, providing 40 % of total crop production (nearly 60 % of cereal production) [61]. Although irrigated agriculture is expected to increase to provide 47 % of global crop production by 2030, rainfed agriculture will remain an important contributor to global food production [61, 62]. Any precipitation change, especially changes in seasonal precipitation, will thus greatly influence the magnitude and direction of climate impacts on water availability, crop production, and food security [63]. Although projections of precipitation change in climate models depend on the relative rates of warming in all regions, a general increase in precipitation is expected in high latitudes, especially in winter, and an overall decrease in many parts of the tropics and subtropics [66].

Agricultural productivity and food security worldwide are thus expected to be strongly affected by climate change in the coming decades, with higher risk and vulnerability of crop production related to higher temperatures [64], water supply, and water availability [60]. Although, as reported in a recent review [63], greater risks for food security might be posed by changes in year-to-year climate variability and extreme weather events, such as extreme temperatures, drought, heavy rainfall, and flooding. Climate change is expected to increase the incidence and severity of crop pests and diseases [59], and might also have a significant effect on mycotoxins [58].

Mycotoxigenic fungi can infect cereals in the field and during storage. In the field, the most frequent fungal species belong to the genus *Fusarium*. *Aspergillus* species can infect plants in the field, especially if crops suffer drought stress or insect attack, but usually develop, as *Penicillium* species, postharvest under poor storage conditions. A large number of *Fusarium* species are important pathogens for nearly all economically important plant species.

Of particular relevance are cereals, the world's most important source of food and feed, which occupy more than half of the world's harvested area [57], and are common hosts for *Fusarium* and *Aspergillus*. For 2014, the world cereal production was forecast at 2525 million tons, the forecast world cereal utilization was put at 2416 million tons, of which 1094 million tons were destined for food and 851 million tons for feed use, and the world cereal trade was forecast at 356 million tons, all figures higher than those reached in 2013 [56].

Improved sampling and analytical methods for mycotoxins should be developed to determine any climate change effects on the mycotoxin contamination of food and feeds [58].

8.3 Detection methods

The presence of mycotoxins in food and feedstuffs can be checked by fast and simple methods, employed as rapid screening tools, and with sophisticated analytical methods useful for identification and quantification of the amount of each toxin. A detection method should be sensitive, robust, reproducible, and applicable to a wide range of compounds. The screening methods used to detect the presence or absence of toxins are rapid, economic and simple, no skilled personnel is necessary and limited or no sample treatment is required. In contrast, analytical methods are expensive and time-consuming and often require sample pre-treatment. These methods are usually based on liquid chromatography coupled with different detectors, therefore trained personnel is required. In addition to the detection method, the collection of representative samples and the sample extraction protocol are critical issues in obtaining accurate results.

8.3.1 Sampling procedures

As already mentioned, mycotoxins are heterogeneously distributed in food lots with large particle size such as dried figs, groundnuts, or fruits. The distribution of mycotoxins in processed products is generally less heterogeneous than in unprocessed food. The sampling procedure should therefore be different in each case. Taking this into account, the European Commission has established methods of sampling for official controls of the levels of mycotoxins in foodstuffs [74, 75]. These official controls shall be in accordance with provisions of regulations about animal feed and food laws, animal health and animal welfare regulations, the maximum limits for certain mycotoxins, and the sampling criteria for the control of levels of these compounds [53–55, 75]. Depending on the sample, the size of the lot, on volume and weight, the number of incremental samples and the volume or weight of aggregate samples should be different. All of these parameters have recently been clearly established and the variability in results due to the sampling procedure can almost be eliminated [74].

8.3.2 Extraction procedures

The presence of mycotoxins is often analyzed in many different matrices, from food and feedstuffs to processed products. In some cases, therefore, other substances such as salts, proteins, oils, lipids, or sugars can interfere with the analysis. In addition, the amount of toxins is sometimes small and can be masked by interfering compounds. For these reasons, extraction and several cleanup steps are sometimes necessary before mycotoxin analysis.

The initial extraction of toxins from solid matrices usually consists of a solid-liquid extraction, employing organic solvents such as methanol or acetonitrile mixed with water in different proportions and an acidic compound like acetic acid or formic acid [52, 102, 103]. Polar molecules such as fumonisins and hydrophobic compounds such as aflatoxins can thus be extracted. Depending on matrix interferences, different operations can be performed after this. In the case of non-processed products, such as wheat or maize, a centrifugation/filtration step might be enough and the extract can be directly analyzed [52, 103]. In other cases, the solid-liquid extraction is followed by a cleanup step using different strategies such as immunoaffinity columns (IACs; [98–101]). These columns are toxin-selective and useful when only analyzing one group of toxins. However, the availability of antibodies is limited and therefore these columns are not appropriate for performing multi-toxin analysis. In addition, IACs can only be used once, in consequence are expensive, and some mycotoxins can be underestimated after this cleanup [97].

Other solid phase extraction (SPE) or in-tube solid-phase microextractions (SPME) are also cleanup procedures used to avoid matrix effect [92–96]. SPE cartridges are less expensive and less selective than IACs and are often used for mycotoxin analysis. Many different column packings have been described for SPE cleanup, such as silica-gel, C_{18}, florisil or ion-exchange materials, as well as polymers and synthetic peptides [70, 71]. Further, other solid extraction procedures, such as selective or dispersive matrix solid-phase have also been developed to clean and prepare samples from different matrices [89–91]. However, mycotoxins have different chemical properties and universal cartridges are not available when multitoxin analysis is necessary.

Based on all these extractions and from technology developed for pesticide analysis of food and feedstuffs, the QuEChERS methodology (from Quick, Easy, Cheap, Effective, Rugged and Safe) has been used to extract and clean samples for mycotoxin analysis in different matrices [84–88, 102]. This methodology involves first an extraction step based on partitioning with salt-out extraction, and then a dispersive matrix solid-phase step. In this way, excess water, organic acids, fatty acids, lipids, sugar, or pigments are eliminated. However, reduced mycotoxin recovery has been reported [102].

Liquid-liquid partitions and dispersive liquid-liquid extractions have also been used, either to extract and clean liquid samples, or to clean extracts from solid samples [81–83]. Different solvents such as methanol, acetonitrile, acetone, ethyl acetate, chloroform, diethyl ether or mixtures are selected for liquid-liquid extractions. In general, polar solvents and pH play a key role during the extraction process. Two immiscible liquid phases are used to clean samples. For this purpose, solvents such as hexane or cyclohexane are used to eliminate nonpolar contaminants like lipids or cholesterol. Supercritical fluids have also been used to extract aflatoxins and zearalenone derivatives from different matrices [25, 78–80].

Using one or more of the above-mentioned procedures can therefore eliminate the matrix effect and the final extract can be directly analyzed or concentrated before analysis when toxins are present in very low concentrations. These methods can be used under analyst criteria but should guarantee high toxin recovery in order to avoid toxin underestimation.

8.3.3 Mycotoxin analysis

Several detection methods can be used after extraction and cleanup to check for the presence of mycotoxins in samples. The analytical method should be simple, rapid, robust, accurate, highly sensitive, and able to detect a wide range of compounds simultaneously. In addition, high throughput of analyzed samples is also important when a method is selected. Chromatographic methods are often used to analyze the presence of mycotoxins, although ELISAs (enzyme linked immune–sorbent assays) or biosensors have also been applied.

8.3.3.1 Thin layer chromatography (TLC)

Thirteen methods were adopted as official AOAC methods for aflatoxins in different matrices by AOAC International from 1970 to 1980, most based on TLC [24]. TLC is considered useful as a screening method for both semiquantitative and quantitative purposes. This planar chromatography has been an official AOAC method for identification and quantification of mycotoxins at 1 ng/g level since 1990 [23]. In the stationary phase, silica, alumina or cellulose, is immobilized over a glass or plastic plate and different solvents are used as mobile phase. Both 1- and 2-dimensional analyses are frequently used in order to improve resolution. After chromatographic development, the plate is visualized with ultraviolet (UV) or fluorescence detectors. This is a powerful, rapid, and low-cost separation method for detecting the presence of toxins but not the exact amount. High-performance TLC and overpressured layer chromatography techniques are modifications of TLC used for quantitative determination of mycotoxins [20–22].

8.3.3.2 Immunoassay techniques

There are several ELISA kits for detection of mycotoxins. These assays are based on a competitive direct or indirect assay using a primary antibody specific for a toxin, a conjugate of an enzyme and the toxin, or secondary antibodies [18, 19]. This technique, the same as IACs, is useful for detection of a toxin or a family of toxins; the simultaneous detection of multiple toxins was thus developed with a magnetoresistive-based immunoassay with a low detection limit [17]. Although ELISA shows high specificity, cross-reactivity with secondary antibodies may occur, resulting in non-specific signals, results should thus sometimes be confirmed by other analytical methods [70].

8.3.3.3 Bioassays

Bioassays are an alternative to chemical analysis where the toxicity or activity of mycotoxins over cellular lines is checked. Qualitative and quantitative detection can be performed in this way [13–16]. These are highly sensitive and low-cost assays. The receptor-assays recently developed to detect fungal toxins are based on the activity over cells [12].

8.3.3.4 Biosensor technique

The immobilization of molecules such as antibodies, receptors or aptamers over sensor surfaces allows the development of highly sensitive detection methods for mycotoxins. The surfaces have different natures depending on the equipment and the molecule to be immobilized. Several biosensor assays have been developed in this way [8–11]

8.3.3.5 High performance liquid chromatography (HPLC)

HPLC and the newly developed ultrahigh pressure liquid chromatography (UPLC) are the main chromatography techniques used to separate and identify mycotoxins [23]. The toxins can easily be separated with this technique, using different columns and elution mixtures and gradients, and then identified by UV and/or fluorescence. A number of toxins already have natural fluorescence and others can be chemically derivatized. HPLC has thus been a standard method for detection of multiple mycotoxins in several matrices with a wide range of detection limits. This methodology has recently been improved coupled with mass detection (MS). The selectivity of MS detection increases the analysis capacity, including samples without cleanup process, and also enlarges the sensitivity, 50 times greater than fluorescence detection [70]. A number of protocols, impossible to summarize in this chapter, have so far been published for multitoxin analysis using MS detection and various techniques and equipment (atmospheric pressure chemical ionization (APCI), electrospray ionization (ESI), MS-MS detection, ion trap, time of flight (TOF), etc.; [6, 7, 23]). MS detection is accurate and specific and problems derived from compound interference within HPLC separation are avoided. Complex and expensive laboratory equipment is required, however, as is skilled personnel. In addition, high quality mycotoxin standards are needed. Although standards are available for the majority of regulated toxins, certified reference materials (CRM) are necessary to definitively identify and quantify mycotoxins, and this grade of purity is not always available. In addition, new toxic compounds will not be detected, since it is necessary to know in advance which molecules will be checked in MS detection. MS techniques are essential, however, and regularly used in food safety and environmental monitoring.

8.3.3.6 Gas chromatography

Gas chromatography can be applied to identify and quantify the presence of mycotoxins using MS, ionization, or spectroscopy detection. In this case the toxins should be volatile or converted into volatile derivatives. This technique is expensive and not regularly used although many references are available [23].

8.3.3.7 Other analytical methods

Capillary electrophoresis, rapid colorimetric and fluorescent test, fluorescence polarization or competitive lateral flow assays have also been developed to detect mycotoxins [70, 71].

8.3.4 Requirements for mycotoxin analysis methods

As previously mentioned, there are no official methods for mycotoxin analysis. However, the methods in Europe must comply with European Community regulations on screening and confirmatory methods for detection of these compounds [74]. In this sense, methods of analysis should be characterized by the criteria of accuracy, applicability (matrix and concentration range), limit of detection, limit of determination, precision, repeatability, reproducibility, recovery, selectivity, linearity, and measurement of uncertainty. The precision values should be obtained with internationally recognized protocols (ISO or IUPAC international harmonized protocols), and the repeatability and reproducibility should be expressed in an internationally recognized form [55]. In addition to these general requirements, mycotoxin analysis methods must comply with different requirements for confirmatory or semi-quantitative screening methods. The procedures for validation of screening methods by means of an inter-laboratory validation, the verification of the performance of a method validated by means of an inter-laboratory exercise and the single-laboratory validation of a screening method are fixed [74]. In terms of performance criteria, full validation is recommended for confirmatory methods in relevant matrices. The method should be validated in-house including CRMs according to performance criteria for each toxin. Semi-quantitative screening methods are used to select samples with levels of toxins which exceed the concentration of interest. The result in this case will be negative or suspect and should later be verified with a confirmatory method. These methods need inter-laboratory or single validation for each individual mycotoxin and commodity. When new mycotoxins are added to the scope of an existing screening method, a full validation is required to demonstrate the suitability of the method. Continuous method verification must be programmed after the initial validation [74]. In summary, performance criteria for mycotoxin screening methods are used to guarantee comparable results to be applied for public health risk assessment and new legislation.

The Joint Research Centre (JRC) of the European Commission has hosted four European Union Reference Laboratories (EURLs) in the area of food safety control since 2006. The JRC's IRMM (Institute for Reference Materials and Measurements) is the European Union Reference Laboratory (EURL) for mycotoxins. The EURL for mycotoxins coordinates a network of national reference laboratories (NRLs) to facilitate the implementation of European legislation on mycotoxins in food and feed and to obtain high quality results by development and validation of methods, organization of comparative testing, and training of laboratory staff. The activities of the EURL with regard to mycotoxins currently concern aflatoxin B_1, total aflatoxins, ochratoxin A, patulin, deoxynivalenol, zearalenone, fumonisins B_1 and B_2, T-2 toxin, HT-2 toxin, and ergot alkaloids. In order to meet demands for certified reference materials, the JRC joined the German Federal Institute for Materials Research and Testing (BAM) and the British company LGC Ltd. to establish a high quality brand of CRMs in Europe, the European Reference Material (ERM®), which was launched in 2004. The 2015 JRC catalog provides 16 certified reference materials for mycotoxins (Tab. 8.1), of which 11 are aflatoxins.

8.4 International mycotoxin regulations

Epidemics of mycotoxicosis (acute or chronic diseases of humans and animals caused by exposure to mycotoxins) have severely affected human populations over whole regions or several countries throughout history, often in connection with times of famine [5]. In Europe, for example, several large-scale outbreaks of ergotism, a mycotoxicosis caused by toxins produced by the ascomycete *Claviceps purpurea* on rye, killed thousands of people during the Middle Ages, and small epidemics of the disease occurred after the Renaissance. Another mycotoxicosis epidemic took place in some regions of the USSR between 1942 and 1948, when thousands of people died from a hemorraghic disease associated with consumption of bread and other cereal products made from grains left in fields over winter and infected by *Fusarium sporotrichoides* and *F. poae*, which produce the toxin T-2, one of the most acutely toxic trichothecenes [4].This mycotoxicosis is now called alimentary toxic aleukia (ATA). Some severe mycotoxicoses have also affected animals in the 20th century. More than 5000 horses died of equine leukoencephalomalacia (ELEM) in 1934 and 1935 in the United States Midwest after consuming maize infected with *F. verticillioides* (syn. *F. moniliforme*). More than 50 years later, in 1989, *F. verticillioides* was found to produce a toxic metabolite, named fumonisin B_1, which was shown to elicit typical ELEM symptoms in horses and to cause liver cancer in rats and lung edema in pigs [3].

Feeding moldy grain and foods has thus been known to be harmful to humans since at least the Middle Ages. However, the fungal compounds associated with the diseases were not identified until the second half of the 20th century. It was not until the early 1960s that the study of mycotoxins began, triggered by the outbreak of

Tab. 8.1: Certified reference materials for mycotoxins available from the 2015 JRC catalog.

Reference material code	Mycotoxin	Material/matrix	Certified value µg/kg
ERM-BD282	Aflatoxin M_1	Whole milk powder (zero level)	< 0.02
ERM-BD283	Aflatoxin M_1	Whole milk powder (low level)	0.68 ± 0.1
ERM-BD284	Aflatoxin M_1	Whole milk powder (high level)	0.68 ± 0.1
BCR-262R	Aflatoxin B_1	Defatted peanut meal (blank)	< 3
BCR-263R	Aflatoxin B_1 Aflatoxin B_2 Aflatoxin G_1	Defatted peanut meal (medium level)	1.7 ± 2.4 3.0 ± 2.4 3.0 ± 0.5
BCR-264	Aflatoxin B_1	Defatted peanut meal (high level)	206 ± 13
BCR-401R	Aflatoxin B_1 Aflatoxin B_2 Aflatoxin G_1 Aflatoxin G_2	Peanut butter (very low level)	< 0.2 < 0.2 < 0.2 < 0.2
BCR-385R	Aflatoxin B_1 Aflatoxin B_2 Aflatoxin G_1 Aflatoxin G_2 Total aflatoxins	Peanut butter (low level)	1.77 ± 0.30 0.48 ± 0.08 0.9 ± 0.4 0.30 ± 012 3.5 ± 0.5
BCR-375	Aflatoxin B_1	Compound feed (very low level blank)	< 1
ERM-BE375	Aflatoxin B_1 Aflatoxin B_2 Aflatoxin G_1 Aflatoxin G_2	Compound feed (very low level)	2.6 ± 0.4 0.2 ± 0.04 0.4 ± 0.1 < 2
ERM-BE376	Aflatoxin B_1 Aflatoxin B_2 Aflatoxin G_1	Compound feed (high level)	12.9 ± 1.8 0.68 ± 0.1 5.2 ± 0.8
BCR-471	Ochratoxin A	Wheat (blank)	< 0.06
ERM-BC716	Zearalenone (ZON)	Maize	< 5
ERM-BC717	Deoxynivalenol (DON) Nivalenol (NIV) Zearalenone (ZON)	Maize	673 ± 187 53 ± 10 83 ± 9
BCR-377	Deoxynivalenol (DON)	Maize flour (very low level blank)	< 0.05 mg/kg
BCR-396	Deoxynivalenol (DON)	Wheat flour (very low level blank)	< 0.05 mg/kg

Turkey-X disease in the United Kingdom in 1960, an epidemic which resulted in the deaths of thousands of turkey poults, ducklings, and chicks fed diets containing peanut meal from Brazil. The toxicity of the meal was discovered to be due to the presence of *Aspergillus flavus*, and the toxic agent was named aflatoxin (from *A. flavus* toxin). Between 1961 and 1963, several researchers showed that aflatoxins could induce acute liver disease in ducklings and liver cancer in rats, while others were able to discover the existence of four major naturally occurring aflatoxins (B_1, B_2, G_1, G_2), to isolate purified aflatoxins from *A. flavus* cultures, and to accomplish structural characterization and synthesis of aflatoxin B_1 [2]. The discovery of aflatoxins led to extensive international research in the following years, to develop analytical methods for their detection and quantification in food. The methodology was used in epidemiological studies carried out from 1968–1985, uncovering the association of aflatoxin ingestion and incidence of hepatocellular carcinoma (HCC) in human populations, and resulting in the classification of aflatoxins as human carcinogens (Group 1) by the International Agency for Research on Cancer (IARC) [2]. Many fungi were screened for toxic secondary compounds in that period, and over 240 fungal metabolites were reported to be toxigenic by 1984.

In 1969, the Food and Drug Administration (FDA) of the United States set an action level for aflatoxins at 20 ppb for all foods, including animal feed, based on the analytical methodology developed and with the aim of restricting aflatoxin exposure to the lowest possible level. This was the first regulation of mycotoxins in the world. After carrying out animal feed studies in the 1970s and 1980s, the FDA revised this action level in 1982, 1989, and 1990 to 300 ppb for aflatoxins in cottonseed meal and 100 ppb for aflatoxins in maize and peanut products intended for use as feed ingredient for beef cattle, swine, and poultry, keeping the level of 20 ppb for human food and all feed intended for immature animals, dairy animals or unknown destinations.

In the decades following the first mycotoxin regulation by FDA, many countries adopted regulations specifying maximum tolerable levels for the most toxic and/or abundant mycotoxins in many food and feed commodities and products, in order to limit exposure to mycotoxins and protect human and animal health. There were 33 countries with known specific mycotoxin regulations for food and/or feed in 1981, 56 in 1987, 77 in 1995, and 100 in 2003 [1].

The first regulation on aflatoxins in the EU was for aflatoxin B_1 in animal feed in 1976, and EU-harmonized regulations for several aflatoxins in human food were established in 1998 [1]. European regulations for aflatoxins were amended and new EU regulations were established for several other mycotoxins in different foodstuffs, including foods for babies and children, in the following years [51, 53, 54]. The progressive introduction of regulatory limits for mycotoxins by the European Commission has been the main driving force for the development of official mycotoxin analysis methods in food and feed [24].

Aflatoxins in food are the most commonly regulated mycotoxins worldwide. Aflatoxin B_1 is the most common food mycotoxin. The most pronounced aflatoxin contamination has usually been found in food crops grown or stored in the warmer agricultural regions of the world. However, global trading of agricultural commodities makes aflatoxin contamination a worldwide problem for human and animal health. All countries have regulatory limits for at least aflatoxin B_1 or the sum of aflatoxins B_1, B_2, G_1, and G_2 in food and/or feed [50].

Depending on the country, specific regulations have been set up for other mycotoxins mainly in human foods: aflatoxin M_1; the trichothecenes deoxynivalenol, diacetoxyscirpenol, T-2 toxin, and HT-2 toxin; the fumonisins B_1, B_2, and B_3; ochratoxin A; zearalenone; patulin; the ergot alkaloids; agaric acid; phomopsins; and sterigmatocystin [50]. There are less recommended mycotoxin levels for animal feed and they are typically higher than for human food [1, 58].

New mycotoxin regulations are continuously being established as scientific and technological progress provides new evidence of risks to human and/or animal health. They concern not only maximum mycotoxin levels in food and feed but also sampling procedures and criteria for analysis methods. A guidance value for the sum of T-2 and HT-2 toxin in cat food was established in 2013 in the EU, given the toxicity of both mycotoxins for cats [49]. The European Commission has recently set maximum levels for citrinin in food supplements based on rice fermented with red yeast *Monascus purpureus*, due to the nephrotoxicity of this mycotoxin [47, 48]. In addition, methods for sampling of large lots, spices, and food supplements have been regulated, performance criteria for T-2 and HT-2 updated, and criteria for citrinin performance and mycotoxin screening methods have been established [74]. Current regulations are based on scientific opinions of authoritative bodies, such as the Joint FAO/WHO Expert Committee on Food Additives (JECFA), *Codex Alimentarius*, and the European Food Safety Authority (EFSA), which regularly assess the risk of mycotoxins and advise on controls to reduce consumer exposure.

The criteria used as a basis for deciding limits have not been harmonized worldwide and vary in different countries. The most common criteria are: scientific criteria related to data from surveys, toxicological data, analytical methods and distribution of mycotoxins in contaminated commodities [46]. Harmonization of mycotoxin regulations in food and feed have taken place within groups of countries: the European Union (EU-28), MERCOSUR (Mercado Común del Sur, i.e. Common Market of South America), Australia/New Zealand, ASEAN (Association of South East Asian Nations), GCC (Gulf Cooperation Council), and COMESA (Common Market of Eastern and Southern Africa) [58]. However, different mycotoxin limits exist among these groups [50].

Although most food consumed worldwide is grown locally, global food and feed trade increased greatly between 2000 and 2010, affecting cereals, fruit and vegetables, oilseeds, coffee, tea, cocoa, and spices [57]. Mycotoxin contamination of exported food and feeds above legislated maximum levels resulted in trade disruptions with potentially important economic implications [77].

The EU has applied a strict control system to keep out food and feedstuffs with mycotoxin levels above those permitted. A rapid alert system for food and feed

(RASFF) has been in place since 1992 which enables rapid exchange of information between food and feed control authorities in member states and the European Commission, thus easing the coordination of responses to food safety threats, including mycotoxins. Imported food and feed consignments are tested for mycotoxins and other possible hazards, depending on the material, at the external borders of the EU (and the European Economic Area – EEA), and are rejected if they pose a risk to human and animal health. RASFF notification of border rejections are sent to all EEA border posts to ensure that the rejected products do not re-enter the EU through another border post.

Alert and information notifications on mycotoxins were included in the 'Chemical' risk group in RASFF reports until 2002 and thereafter as a separate identified risk. Most RASFF notifications of mycotoxin contamination have concerned aflatoxins, and the European Commission has been establishing new regulations for the protection of public health. The RASFF received a very high number of aflatoxin notifications concerning pistachios from Iran and also for peanuts and hazelnuts from 2003 to 2007. The Council Directive 2005/85/EC was established on January 26, 2005 by the European Commission (EC) and imposed special conditions on pistachios and certain products derived from pistachios originating in and or consigned from Iran. As a result fewer notifications occurred in 2006, although import quantities remained about the same [45]. Further Commission decisions came into force in 2006 and 2007 related to the control of aflatoxins in foodstuffs imported from Egypt, China, Turkey, Brazil, and Iran (Commission Decision 2006/504/EC of July 12, 2006), in almonds and derived products from the United States (Commission Decision 2007/5063/EC of August 1, 2007), and in peanuts and derived products from Brazil (Commission Decision 2007/759/EC of November 19, 2007). All of these regulations have resulted in a decreasing number of aflatoxin notifications from 2003 to 2007 [44].

In 2008, however, there was a 28 % increase in aflatoxin notifications and 23 % for mycotoxins in general. The increase was noted in the product categories 'Nuts, nut products and seeds' and 'Cereal products', although no single cause for this could be identified [43].

In the period 2008–2014, there were 9338 EU border rejection notifications from third countries for food, of which 3174 (34 %) were due to mycotoxins [37–43]; see Fig. 8.1.

Most rejections were of nuts and nut products imported from China, Turkey, and Iran, and to a lesser extent from the United States and Argentina, but also remarkable was a record of 480 rejections of figs (classified in the category 'Fruits and vegetables') from Turkey, and 202 in the category 'Herbs and spices' from India; see Fig. 8.2.

As in the previous years, most mycotoxin rejections were due to elevated levels of aflatoxins (Fig. 8.3) according to the EU aflatoxin regulations (European Commission 2006, 2010). The number of border rejections of food due to aflatoxin contamination greatly decreased from 2008 to 2014, which might be related to changes in EU aflatoxin regulations. According to RASFF data, the frequency of control of the import of certain foodstuffs from certain third countries due to contamination risk by aflatoxin was increased with the adoption of Commission Regulation (EC) No 1152/2009 of November 27, 2009. In February 2010 Commission Regulation (EU) No 165/2010 amended the in force Regulation (EC) no 1881/2006 increasing maximum levels for aflatoxins in almonds, hazelnuts, pistachios, and other tree nuts, taking into account developments in *Codex Alimentarius* and new scientific information.

The first RASFF notifications of aflatoxins in different feed materials (only two) were reported in 2005 [36]. There were 618 EU border rejection notifications from third countries for feed in the period 2008–2014, and 235 (38 %) were of feed materials and pet food (Fig. 8.4) contaminated with aflatoxins [37–43].

The number of border rejections peaked in 2011 and 2012, and there were fewer EU aflatoxin rejections of feed products in 2013 and 2014 (Fig. 8.5). Most of the contaminated feedstuffs (72 %) were groundnuts and groundnut kernels from India for bird feed. The Indian authorities presented an action plan to address the issue in 2011 on request of the European Commission.

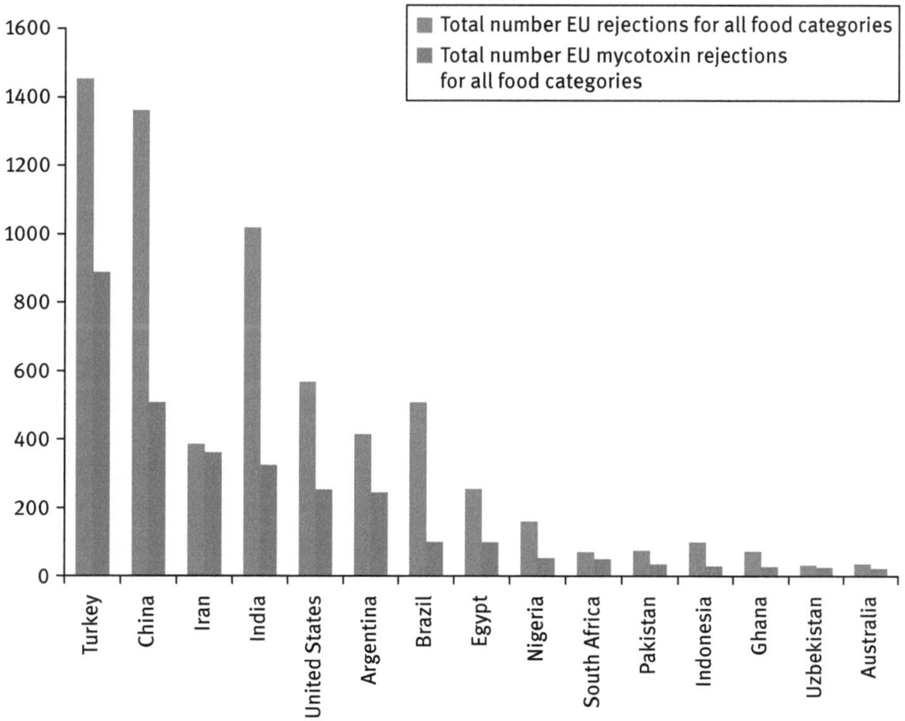

Fig. 8.1: Number of EU border rejections (total and mycotoxin) from third countries for food, 2008–2014. Data extracted from information published by the EU-Rapid Alert System for Foods and Feeds (RASFF).

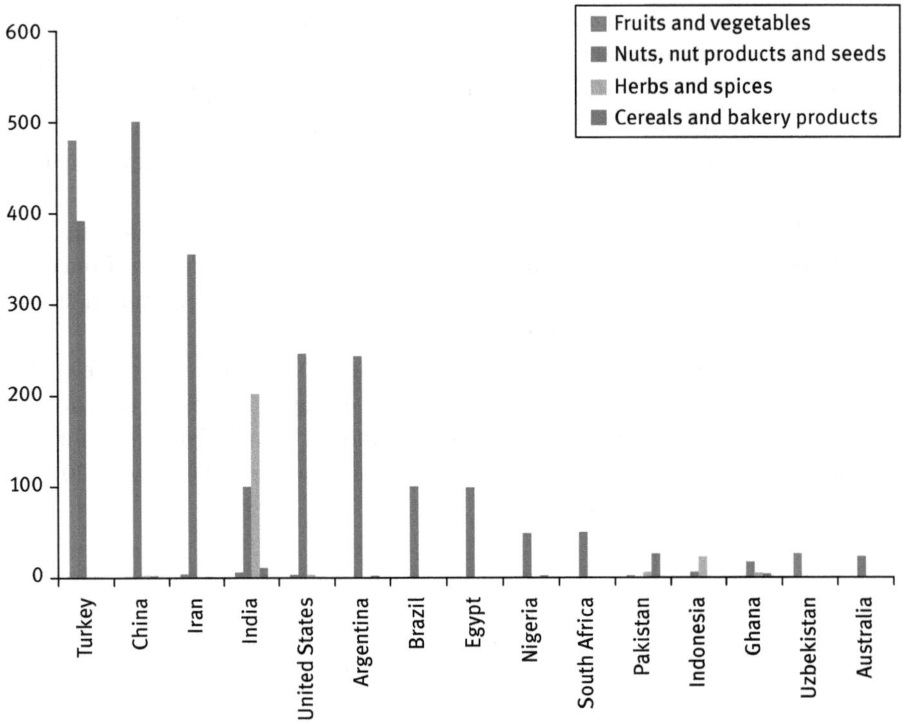

Fig. 8.2: Number of EU mycotoxin border rejections from third countries according to food product category and country of origin, 2008–2014. Data extracted from information published by the EU-Rapid Alert System for Foods and Feeds (RASFF).

Aflatoxin testing is not mandatory for domestically produced or imported foods and feed ingredients in the United States, although the FDA regularly takes samples of food and feed imports [35]. This explains the significant differences between the EU and the US in the pattern of rejections, reflecting both trade patterns and distinct food safety and other requirements, as well as the associated systems of enforcement [34]. The most frequent reason for rejection of food imports is non-compliance with labelling in the US (Fig. 8.6), being associated with 58 % of all rejections. However, the EU does not generally enforce labelling requirements through border inspection.

The food supply chain is increasingly interconnected and the world market faces different national and international food safety regulations and standards which might create an uncertain global regulatory environment. The commercial interests of individual countries and the need for sufficient availability of food have hampered global harmonization of mycotoxin regulations.

The European Union is revising mycotoxin legislation continuously. In 2013, at the request of the European Commission, the EFSA CONTAM Panel delivered a statement regarding the public health risks related to a possible increase in the maximum

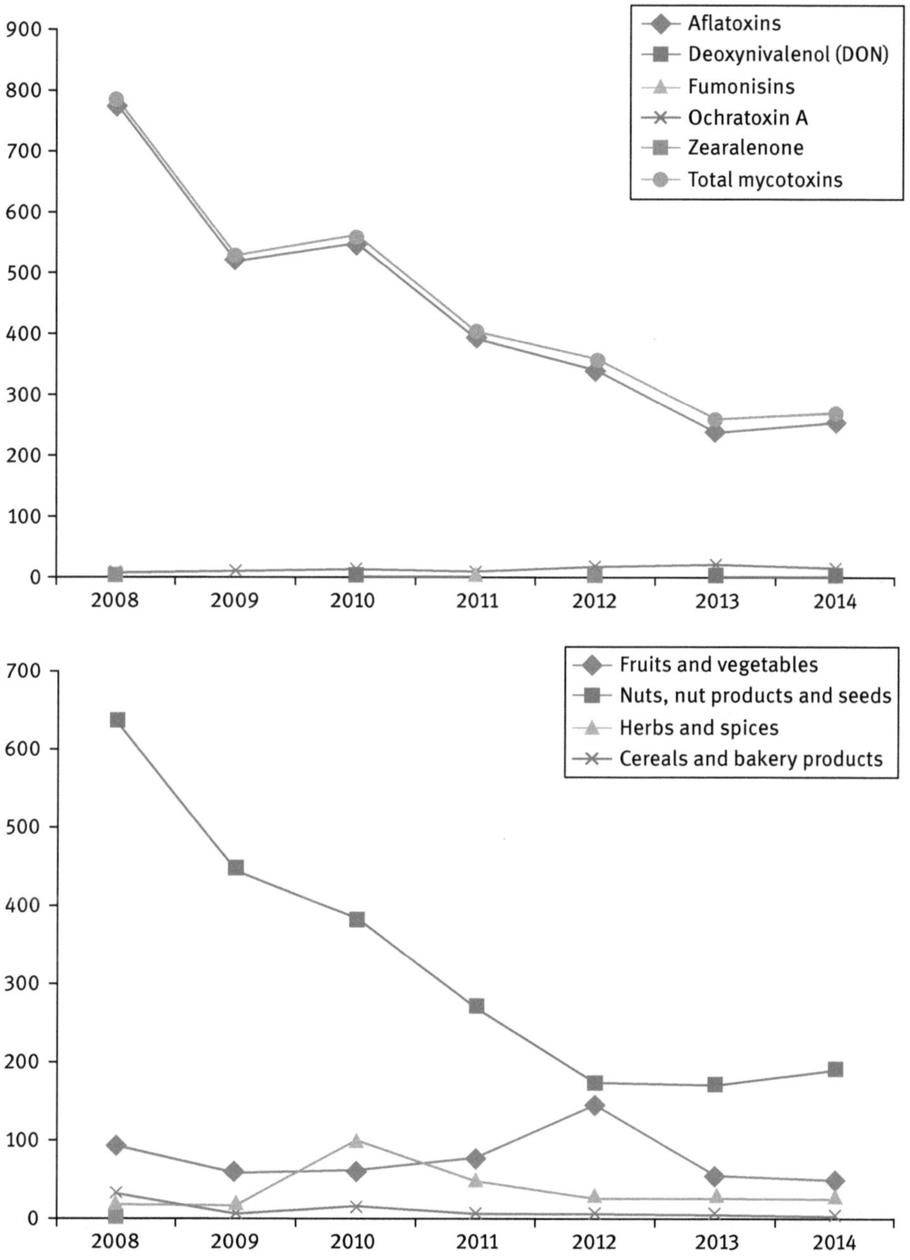

Fig. 8.3: Number of EU mycotoxin border rejections from third countries for food based on legislated mycotoxins and according to food product category, 2008–2014. Data extracted from information published by the EU-Rapid Alert System for Foods and Feeds (RASFF).

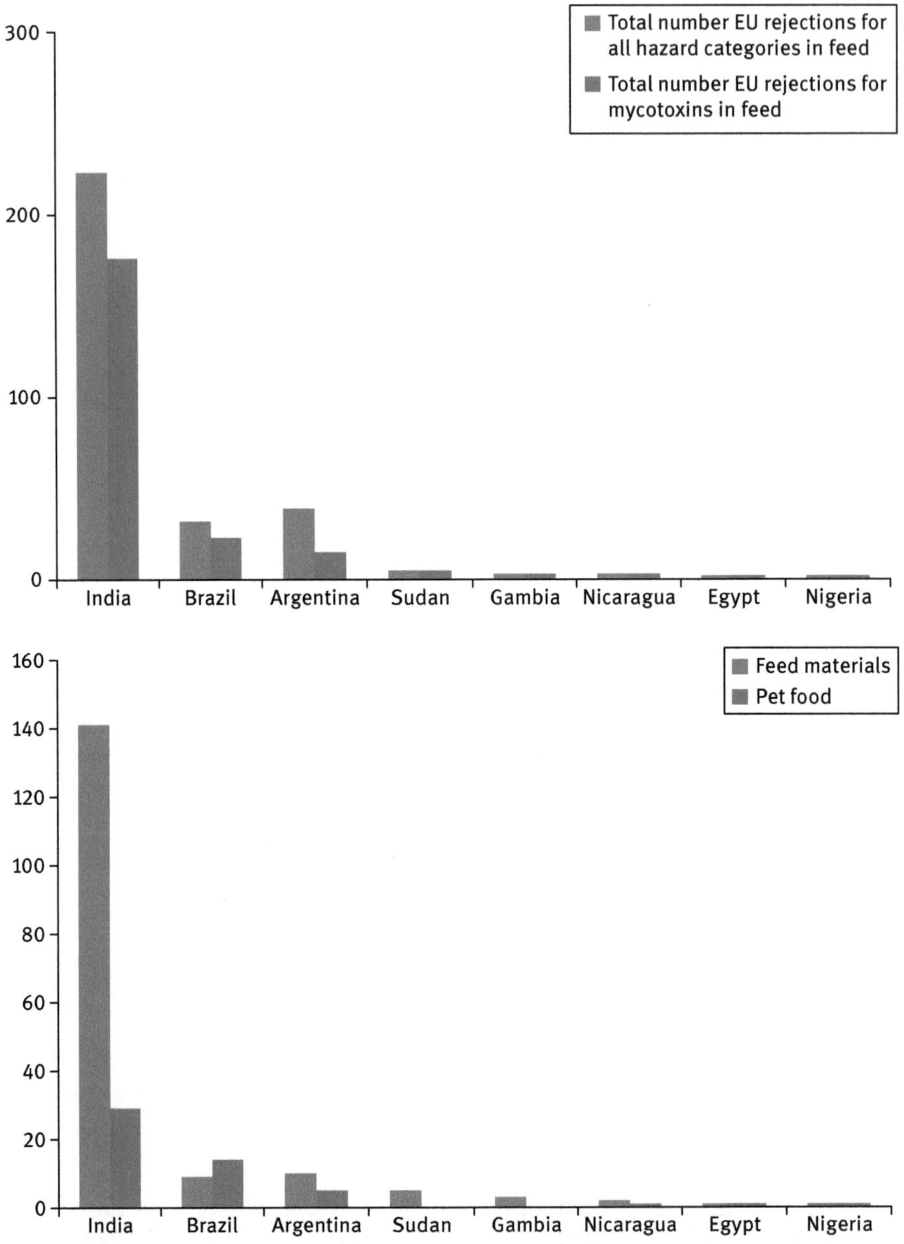

Fig. 8.4: Number of EU border rejections (total and mycotoxin) from third countries for feed, and according to feed category, 2008–2014. Data extracted from information published by the EU-Rapid Alert System for Foods and Feeds (RASFF).

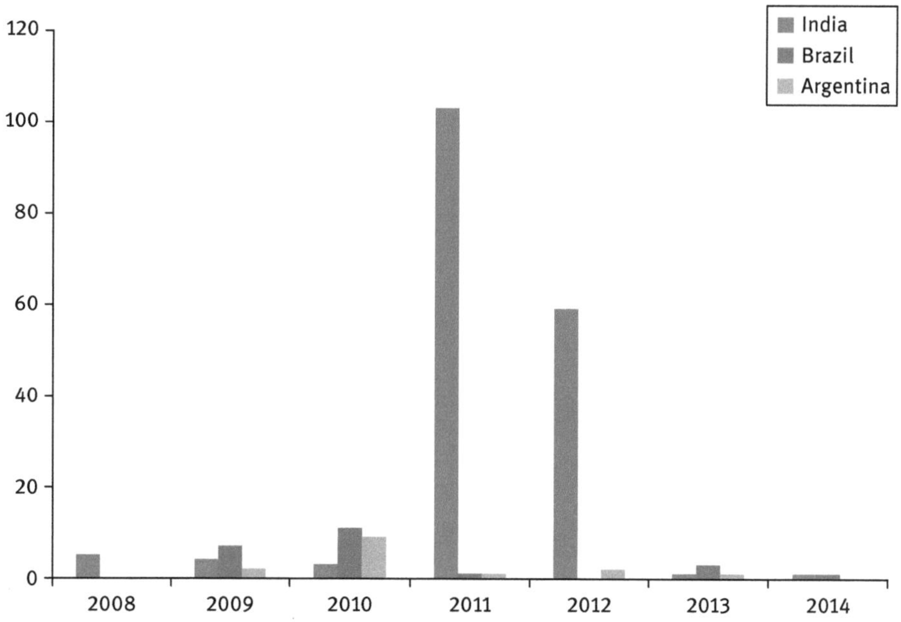

Fig. 8.5: Number of EU mycotoxin border rejections from third countries for feed according to country of origin, 2008–2014. Data extracted from information published by the EU-Rapid Alert System for Foods and Feeds (RASFF).

level of DON for certain semi-processed cereal products from 750 µg/kg to 1000 µg/kg [30]. The study concluded that an increase in DON maximum level could be expected to be associated with an increase in DON and its 3- and 15-acetyl-derivatives, therefore increasing exposure of the group health based guidance values. The European Commission also recently requested an urgent scientific statement on the increased public health risk related to a possible temporary derogation from the maximum level (tML) of deoxynivalenol (DON), fumonisins (FUM), and zearalenone (ZON) in maize and maize products. The request referred to the remarkable increase in DON, FUM, and ZON contamination in maize and maize products from the 2013 harvest in France which led this country to request a temporary derogation of the maximum levels for those mycotoxins. The EFSA concluded that, together with the uptake of other food commodities, the detected increase in mycotoxin contamination was already close to what is considered a safe level for some consumers, and an extended study on maize mycotoxins from the 2013 harvest in other European production regions was necessary [28].

Regulations should be optimized to protect human and animal health, and at the same time not obstruct the free trade market [33].

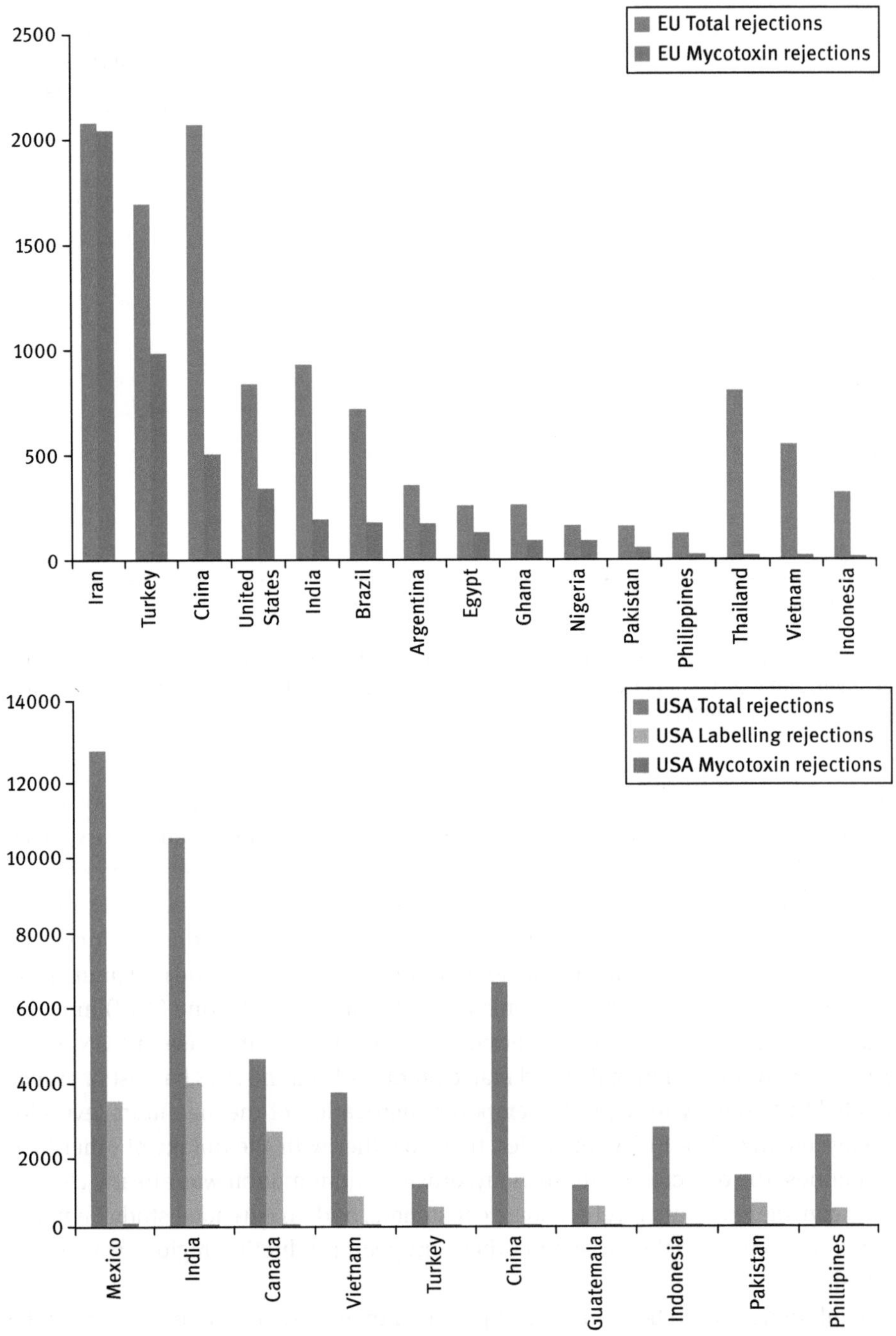

Fig. 8.6: Number of EU rejections of food and feed products and of USA rejections of food products, from third countries, 2002–2008. Data extracted from information published by the United Nations Industrial Development Organization (UNIDO) [34].

8.5 Mycotoxin legislation and climate change

Climate change is expected to increase the incidence and severity of pests and diseases in crops [59], and might have a significant effect on mycotoxins and their regulation [58].

In 2012, there were five notifications of aflatoxin M_1 in milk at levels above the maximum level of 0.05 ppb imposed by EU legislation. All were related to an increased prevalence of aflatoxins in maize for animal feed from southeast Europe [39]. The growing season of maize in that area had been characterized by a very severe drought, making the crop vulnerable to infection by *Aspergillus flavus* and *A. parasiticus* and resulting in increased prevalence of aflatoxins [39].

In the EU, risk of aflatoxin contamination in maize is expected to be higher in some southern European countries, and low to medium in the four main maize producing countries (Romania, France, Hungary, and northeast Italy) in a +2 °C climate change scenario, whilst in a +5 °C scenario, aflatoxin levels are predicted to be lower but risks are expected to be broader and increase towards northern EU countries [32]. These predictions of aflatoxin risks will force a change in the notion that aflatoxins are mostly a problem of imported food and feed in Europe. In a climate change scenario, a strict control system will have to be applied not only to imported food and feedstuffs but also to national agricultural commodities susceptible to contamination by aflatoxins.

Although research on the effects of climate change on mycotoxins has been focused on aflatoxins due to their high toxicity for humans and animals, much attention will have to be paid in the future to the other regulated mycotoxins and to those not yet regulated.

In the last five years the European Food Safety Authority (EFSA) has delivered scientific opinions on the risks for animal and public health related to the presence of nivalenol, T-2 and HT-2 toxin, citrinin, beauvericin and enniatins, and modified forms of the *Fusarium* toxins zearalenone, nivalenol, T-2 and HT-2 toxins and fumonisins in food and feed [27–31]. Modified mycotoxins are metabolites of the parent mycotoxin formed in the plant or fungus, e.g. by conjugation with polar compounds, commonly glucose but also with sulfate and glutathione. Although some conclusions were drawn, the CONTAM panel identified a number of uncertainties and data gaps. The need for evaluation of the analytical methods in inter-laboratory validation studies, development of performance criteria, development of reference materials in food and feed, performance of *in vivo* toxicity studies and human risk assessment, acquisition of more data on farm animal toxicity and the investigation of the carry-over of one of the studied mycotoxins from feed to animal products intended for human consumption was highlighted with regard to some of the mycotoxins studied.

In a further scientific opinion from December 2014 on the risks for human and animal health related to the presence of modified forms of the *Fusarium* toxins zearalenone, nivalenol, T-2 and HT-2 toxins and fumonisins [26], the CONTAM Panel asked for more information on the chemical structures and toxicological data of modified mycotoxins, as well as for further work to identify modified mycotoxins not yet characterized, development of properly validated and sensitive routine analytical methods, and further investigation of the fate of modified forms upon food and feed processing.

Mycotoxin regulations may have an impact not only on international trade of food and feed products but also on food security in developing countries in a climate change scenario, as demand for agricultural products is estimated to increase by about 50 % by 2030 as the world population increases [61].

The interaction between mycotoxigenic fungi and food crops in the field will not only continue but could be extended to newer agricultural areas, so the challenge to limit global exposure to mycotoxins for the protection of human and animal health will rely on scientific and political collaboration both nationally and internationally. As suggested by Van Egmond [58], regulations may force food and feed business operators to apply good agricultural, manufacturing, storage, and hygiene practices with regard to mycotoxins, which might eventually result in reduced occurrences of undesired human exposure. Regular monitoring of fungi and their mycotoxins in the field and in the food and feed-processing companies will be indispensable, but most efforts will need to be focused on the prevention of mycotoxin contamination in the field (biocontrol of mycotoxigenic fungi, control of insects, procurement of resistant cultivars, sustainable irrigation, and the development of predictive models for early stages in crop growth on smaller scales), and on ensuring safe storage conditions for food and feedstuffs.

References

[1] Van Egmond H. Regulations relating to mycotoxins in food. *Anal Bioanal Chem* 2007;389:147–157.

[2] Kensler TW, Roebuck BD, Wogan GN, Groopman JD. Aflatoxin: a 50-year odyssey of mechanistic and translational toxicology. *Toxicological sciences* 2011;120 Suppl 1:S28–48.

[3] Dutton MF. Fumonisins, mycotoxins of increasing importance: their nature and their effects. *Pharmacology & therapeutics* 1996;70:137–161.

[4] Li Y, Wang Z, Beier RC et al. T-2 toxin, a trichothecene mycotoxin: review of toxicity, metabolism, and analytical methods. *Journal of agricultural and food chemistry* 2011;59:3441–3453.

[5] Peraica M, Rasic D. The impact of mycotoxicoses on human history. *Arhiv za higijenu rada i toksikologiju* 2012;63:513–518.

[6] O'Mahony J, Clarke L, Whelan M, O'Kennedy R, Lehotay SJ, Danaher M. The use of ultra-high pressure liquid chromatography with tandem mass spectrometric detection in the analysis of agrochemical residues and mycotoxins in food - challenges and applications. *Journal of chromatography A* 2013;1292:83–95.

[7] Li P, Zhang Z, Hu X, Zhang Q. Advanced hyphenated chromatographic-mass spectrometry in mycotoxin determination: current status and prospects. *Mass spectrometry reviews* 2013;32:420–452.

[8] Xu X, Liu X, Li Y, Ying Y. A simple and rapid optical biosensor for detection of aflatoxin B_1 based on competitive dispersion of gold nanorods. *Biosensors & bioelectronics* 2013;47:361–367.

[9] Park JH, Byun JY, Mun H et al. A regenerable, label-free, localized surface plasmon resonance (LSPR) aptasensor for the detection of ochratoxin A. *Biosensors & bioelectronics* 2014;59:321–327.

[10] Zhu Z, Feng M, Zuo L et al. An aptamer based surface plasmon resonance biosensor for the detection of ochratoxin A in wine and peanut oil. *Biosensors & bioelectronics* 2014;65C: 320–326.

[11] Vidal JC, Bonel L, Ezquerra A et al. Electrochemical affinity biosensors for detection of mycotoxins: A review. *Biosensors & bioelectronics* 2013;49:146–158.

[12] Molina-Molina JM, Real M, Jimenez-Diaz I et al. Assessment of estrogenic and anti-androgenic activities of the mycotoxin zearalenone and its metabolites using in vitro receptor-specific bioassays. *Food and chemical toxicology* 2014;74:233–9.

[13] Clarke R, Connolly L, Frizzell C, Elliott CT. Cytotoxic assessment of the regulated, co-existing mycotoxins aflatoxin B_1, fumonisin B_1 and ochratoxin, in single, binary and tertiary mixtures. *Toxicon* 2014;90:70–81.

[14] Wang SK, Liu S, Yang LG, Shi RF, Sun GJ. Effect of fumonisin B(1) on the cell cycle of normal human liver cells. *Molecular medicine reports* 2013;7:1970–1976.

[15] Moscone D, Arduini F, Amine A. A rapid enzymatic method for aflatoxin B detection. *Methods in molecular biology* 2011;739:217–235.

[16] Abolmaali S, Mitterbauer R, Spadiut O et al. Engineered bakers yeast as a sensitive bioassay indicator organism for the trichothecene toxin deoxynivalenol. *Journal of microbiological methods* 2008;72:306–312.

[17] Mak AC, Osterfeld SJ, Yu H et al. Sensitive giant magnetoresistive-based immunoassay for multiplex mycotoxin detection. *Biosensors & bioelectronics* 2010;25:1635–1639.

[18] Abramson D, Usleber E, Martlbauer E. Rapid determination of citrinin in corn by fluorescence liquid chromatography and enzyme immunoassay. *Journal of AOAC International* 1999;82:1353–1356.

[19] Abramson D, Usleber E, Martlbauer E. An indirect enzyme immunoassay for the mycotoxin citrinin. *Applied and environmental microbiology* 1995;61:2007–2009.

[20] Moricz AM, Ott PG, Billes F, Otta KH, Tyihak E. The influence of L-ascorbic acid on the antibacterial-toxic activity of aflatoxins on adsorbent layer. *Journal of applied microbiology* 2007;103:2525–2532.

[21] Otta KH, Papp E, Bagocsi B. Determination of aflatoxins in food by overpressured-layer chromatography. *Journal of chromatography A* 2000;882:11–16.

[22] Kumagai S, Nakajima M, Tabata S et al. Aflatoxin and ochratoxin A contamination of retail foods and intake of these mycotoxins in Japan. *Food additives & contaminants Part A, Chemistry, analysis, control, exposure & risk assessment* 2008;25:1101–1106.

[23] Rahmani A, Jinap S, Soleimany F. Qualitative and quantitative analysis of mycotoxins. *Comprehensive reviews in food science and food safety* 2009;8:202–251.

[24] Senyuva H. Official methods and performance criteria for determining mycotoxins in food and feed. In: Saeger D (ed). *Determining mycotoxins and mycotoxigenic fungi in food and feed.* Sawston, Cambridge, UK: Woodhead Publishing Limited; 2011:171–193.

[25] Zougagh M, Rios A. Supercritical fluid extraction of macrocyclic lactone mycotoxins in maize flour samples for rapid amperometric screening and alternative liquid chromatographic method for confirmation. *Journal of chromatography A* 2008;1177:50–57.

[26] EFSA. *Scientific Opinion on the risks for human and animal health related to the presence of modified forms of certain mycotoxins in food and feed.* EFSA J 2014a;12:3916.

[27] EFSA. *Scientific Opinion on the risks to human and animal health related to the presence of beauvericin and enniatins in food and feed.* EFSA J 2014b;12:3802.

[28] EFSA. *Evaluation of the increase of risk for public health related to a possible temporary derogation from the maximum level of deoxynivalenol, zearalenone and fumonisins for maize and maize products.* EFSA J 2014c;12:3699.

[29] EFSA. *Scientific Opinion on the risks for animal and public health related to the presence of T-2 and HT-2 toxin in food and feed.* EFSA J 2011;9:2481.

[30] EFSA. *Scientific Opinion on risks for animal and public health related to the presence of nivalenol in food and feed.* EFSA J 2013;11:3262.

[31] EFSA. *Scientific Opinion on the risks for public and animal health related to the presence of citrinin in food and feed.* EFSA J 2012;10:2605.

[32] Battilani P, Rossi V, Giorni P et al. *Modelling, predicting and mapping the emergence of aflatoxins in cereals in the EU due to climate change.* European Food Safety Authority (EFSA) Supporting publication 2012.

[33] Tasheva-Petkova Y, Christova-Bagdassarian V, Tachev A, Atanassova M. Harmonize approaches to analysis and risk assessment of mycotoxins in foodstuffs. *Int J Adv Res* 2014;2:1097–1116.

[34] Henson S, Olale E. What do border rejections tell us about trade standards compliance of developing countries? Analysis of EU and US Data 2002–2008. In. *Slovakia: Secretariat of the United Nations Industrial Development organization(UNIDO) working paper.* 2011.

[35] Dohlman E. Mycotoxin hazards and regulations: Impacts on food and animal feed crop trade. In: Buzby J (ed*). International Trade and Food Safety: Economic theory and case studies.* Washington DC: USDA Agricultural Economic Report; 2003:97–108.

[36] RASFF-EU. RASFF Annual report 2005. Luxemburg; 2005.

[37] RASFF-EU. RASFF Annual report 2014. Luxemburg; 2014.

[38] RASFF-EU. RASFF Annual report 2013. Luxemburg; 2013.

[39] RASFF-EU. RASFF Annual report 2012. Luxemburg; 2012.

[40] RASFF-EU. RASFF Annual report 2011. Luxemburg; 2011.

[41] RASFF-EU. RASFF Annual report 2010. Luxemburg; 2010.

[42] RASFF-EU. RASFF Annual report 2009. Luxemburg; 2009.

[43] RASFF-EU. RASFF Annual report 2008. Luxemburg; 2008.

[44] RASFF-EU. RASFF Annual report 2007. Luxemburg; 2007.

[45] RASFF-EU. RASFF Annual report 2006. Luxemburg; 2006.

[46] van Egmond HP, Dekker WH. Worldwide regulations for mycotoxins in 1994. *Natural toxins* 1995;3:332–336; discussion 341.

[47] (212/2014) CRE. COMMISSION REGULATION (EU) No 212/2014 of 6 March 2014 amending Regulation (EC) No 1881/2006 as regards maximum levels of the contaminant citrinin in food supplements based on rice fermented with red yeast Monascus purpureus. In: Communities CoTE (ed). *Official Journal of the European Union*; 2014:L 67/3–4.

[48] Flajs D, Peraica M. Toxicological properties of citrinin. *Arhiv za higijenu rada i toksikologiju* 2009;60:457–464.

[49] 2013/367/EU CR. COMMISSION RECOMMENDATION of 4 November 2013 amending Recommendation 2006/576/EC as regards T-2 and HT-2 toxin in compound feed for cats. In: Communities CoTE (ed). *Official Journal of the European Union*; 2013:L 294/44.

[50] FAO (ed). *Worldwide regulations for mycotoxins in food and feed in 2003.* Rome, FAO; 2004.

[51] (105/2010) CRE. Commission Regulation (EU) No 105/2010 of 5 February 2010 amending Regulation (EC) No 1881/2006 setting maximum levels for cetain contaminants in foodstuffs as regards ochratoxin A. In: Communities CoTE (ed). *Official Journal of the European Union*; 2010:L 35/7-/8.

[52] Sulyok M, Berthiller F, Krska R, Schuhmacher R. Development and validation of a liquid chromatography/tandem mass spectrometric method for the determination of 39 mycotoxins in wheat and maize. *Rapid communications in mass spectrometry: RCM* 2006;20:2649–59.
[53] (1126/2007) CRE. Commission Regulation (EC) No 1126/2007 of 28 September 2007 amending Regulation (EC) No 1881/2006 setting maximum levels for certain contaminants in foodstuffs as regards Fusarium toxins in maize and maize products. In: Communities CoTE (ed). *Official Journal of the European Union*; 2007:L 255/14-/17.
[54] (1881/2006) CRE. Commission Regulation (EC) No 1881/2006 of 19 December 2006 setting maximum levels for certain contaminants in foodstuffs. In: Communities CoTE (ed). *Official Journal of the European Union*; 2006:L 364/5-/24.
[55] (882/2004) CRE. Regulation (EC) No 882/2004 of the European Parliament and of the Council of 29 April 2004 on official controls performed to ensure the verification of compliance with feed and food law, animal health and animal welfare rules In: Communities CoTE (ed). *Official Journal of the European Union*; 2004:L 165/1-/41.
[56] FAO. *Cereal supply and demand brief*. Rome, FAO; 2014.
[57] FAO. *Statistical year book 2013. World book and agriculture*. Rome, FAO; 2013.
[58] Van Egmond H. Mycotoxins: Risks, regulations and European cooperation. *J Nat sci, Matica Srpska Novi Sad* 2013;125:7–20.
[59] Roos J, Hopkins R, Kvarnheden A, Dixelius C. The impact of global warming on plant diseases and insect vectors in Sweden. *Eur J Plant Pathol* 2011;129:9–19.
[60] FAO. *Climate change, water and food security*. Rome, FAO; 2008.
[61] Bruinsma J (ed). *World agriculture: towards 2015/2030. An FAO perspective*. Rome, FAO and London, UK: Earthscan Publications Ltd; 2003.
[62] FAO. *The state of the world's land and water resources for food and agriculture. Managing systems at risk*. Rome, FAO and London, UK: Earthscan Ltd; 2011.
[63] Gornall J, Betts R, Burke E et al. Implications of climate change for agricultural productivity in the early twenty-first century. Philosophical transactions of the Royal Society of London Series B, *Biological sciences* 2010;365:2973–2989.
[64] Battisti DS, Naylor RL. Historical warnings of future food insecurity with unprecedented seasonal heat. *Science* 2009;323:240–244.
[65] Hansen J, Sato M, Ruedy R, Lo K. Global surface temperature change. *Reviews of Geophysics* 2010;48,RG4004, doi:10.1029/2010RG000345.
[66] IPCC. *Climate change 2007: the physical science basis. Contribution of Working Group I to the Fourth Assessment Report of the Intergovenmental Panel on Climate Change*. Cambridge, UK: Cambridge University Press; 2007.
[67] Osytry V. *Alternaria* mycotoxins: an overview of chemical characterization, producers, toxicity, analysis and occurrence in foodstuffs. *World Mycotoxin Journal* 2008;1:175–188.
[68] Marin S, Ramos AJ, Cano-Sancho G, Sanchis V. Mycotoxins: occurrence, toxicology, and exposure assessment. *Food and chemical toxicology* 2013;60:218–237.
[69] Prandini A, Tansini G, Sigolo S, Filippi L, Laporta M, Piva G. On the occurrence of aflatoxin M_1 in milk and dairy products. *Food and chemical toxicology* 2009;47:984–991.
[70] Yang J, Li J, Jiang Y et al. Natural occurrence, analysis, and prevention of mycotoxins in fruits and their processed products. *Critical reviews in food science and nutrition* 2014;54:64–83.
[71] Anfossi L, Baggiani C, Giovannoli C, Giraudi G. Mycotoxins in food and feed: Extraction, analysis and emerging technologies for rapid and on-field detection. *Recent patents on food, nutrition & agriculture* 2010;2:140–153.
[72] (15/2011) CRE. Commission Regulation (EU) No 15/2011 of 10 January 2011 amending Regulation (EC) No 2074/2005 as regards recognised testing methods for detecting marine biotoxins in live bivalve molluscs. *Official Journal of the European Communities* 2011;L6:3–4.

[73] Botana LM. A perspective on the toxicology of marine toxins. *Chemical research in toxicology* 2012;25:1800–1804.

[74] (519/2014) CRE. Commission Regulation (EU) No 519/2014 of 16 May 2014 amending Regulation (EC) No 401/2006 as regards methods of sampling of large lots, spices and food supplements, performance criteria for T-2, HT-2 toxin and citrinin and screening methods of analysis. In: Communities CoTE (ed). *Official Journal of the European Union*; 2014:L 147/29-/43.

[75] (401/2006) CRE. Commission Regulation (EC) No 401/2006 of 23 February 2006 laying down the methods of sampling and analysis for the official control of the levels of mycotoxins in foodstuffs. In: Communities CoTE (ed). *Official Journal of the European Union*; 2006:L 70/12–70/34.

[76] Smith JE, Lewis CW, Anderson JG, Solomon GL. *Mycotoxins in Human Nutrition and Health*. Directorate-General XII Science, Research and Development. EUR 16048 EN 1994.

[77] Tirado M, Clarke R, Jaykus L, McQuatters-Gollop A, Frank J. Climate change and food safety: A review. *Food Research International* 2010;43:1745–1765.

[78] Ehlers D, Czech E, Quirin KW, Weber R. Distribution of aflatoxins between extract and extraction residue of paprika using supercritical carbon dioxide. *Phytochem Anal* 2006;17:114–120.

[79] Starr JM, Selim MI. Supercritical fluid extraction of aflatoxin B(1) from soil. *Journal of chromatography A* 2008;1209:37–43.

[80] Liau BC, Jong TT, Lee MR, Chang CM. Supercritical fluid extraction and quantification of aflatoxins in Zizyphi Fructus by liquid chromatography/atmospheric pressure chemical ionization tandem mass spectrometry. *Rapid communications in mass spectrometry* 2007;21: 667–673.

[81] Arroyo-Manzanares N, Huertas-Perez JF, Gamiz-Gracia L, Garcia-Campana AM. A new approach in sample treatment combined with UHPLC-MS/MS for the determination of multiclass mycotoxins in edible nuts and seeds. *Talanta* 2013a;115:61–67.

[82] Shephard GS, Marasas WF, Leggott NL, Yazdanpanah H, Rahimian H, Safavi N. Natural occurrence of fumonisins in corn from Iran. *Journal of agricultural and food chemistry* 2000;48:1860–1864.

[83] Sorensen LK, Elbaek TH. Determination of mycotoxins in bovine milk by liquid chromatography tandem mass spectrometry. *J Chromatogr B* 2005;820:183–196.

[84] Liu Y, Han S, Lu M, Wang P, Han J, Wang J. Modified QuEChERS method combined with ultra-high performance liquid chromatography tandem mass spectrometry for the simultaneous determination of 26 mycotoxins in sesame butter. *J Chromatogr B* 2014;970:68–76.

[85] Zhang JM, Wu YL, YB. L. Simultaneous determination of carbamate insecticides and mycotoxins in cereals by reversed phase liquid chromatography tandem mass spectrometry using a quick, easy, cheap, effective, rugged and safe extraction procedure. *J Chromatogr B* 2013;915–916:13–20.

[86] Arroyo-Manzanares N, Garcia-Campana AM, Gamiz-Gracia L. Comparison of different sample treatments for the analysis of ochratoxin A in wine by capillary HPLC with laser-induced fluorescence detection. *Anal Bioanal Chem* 2011;401:2987–2994.

[87] Rasmussen RR, Storm IM, Rasmussen PH, Smedsgaard J, Nielsen KF. Multi-mycotoxin analysis of maize silage by LC-MS/MS. *Anal Bioanal Chem* 2010;397:765–776.

[88] Zachariasova M, Lacina O, Malachova A et al. Novel approaches in analysis of Fusarium mycotoxins in cereals employing ultra performance liquid chromatography coupled with high resolution mass spectrometry. *Analytica chimica acta* 2010;662:51–61.

[89] Cavaliere C, Foglia P, Guarino C, Nazzari M, Samperi R, Lagana A. Determination of aflatoxins in olive oil by liquid chromatography-tandem mass spectrometry. *Analytica chimica acta* 2007;596:141–148.

[90] Blesa J, Soriano JM, Molto JC, Marin R, Manes J. Determination of aflatoxins in peanuts by matrix solid-phase dispersion and liquid chromatography. *Journal of chromatography A* 2003;1011:49–54.

[91] Bacaloni A, Cavaliere C, Cucci F, Foglia P, Samperi R, Lagana A. Determination of aflatoxins in hazelnuts by various sample preparation methods and liquid chromatography-tandem mass spectrometry. *Journal of chromatography A* 2008;1179:182–189.

[92] Nonaka Y, Saito K, Hanioka N, Narimatsu S, Kataoka H. Determination of aflatoxins in food samples by automated on-line in-tube solid-phase microextraction coupled with liquid chromatography-mass spectrometry. *Journal of chromatography A* 2009;1216:4416–4422.

[93] Vosough M, Bayat M, Salemi A. Matrix-free analysis of aflatoxins in pistachio nuts using parallel factor modeling of liquid chromatography diode-array detection data. *Analytica chimica acta* 2010;663:11–18.

[94] Saito K, Ikeuchi R, Kataoka H. Determination of ochratoxins in nuts and grain samples by in-tube solid-phase microextraction coupled with liquid chromatography-mass spectrometry. *Journal of chromatography A* 2012;1220:1–6.

[95] Fu Z, Huang X, Min S. Rapid determination of aflatoxins in corn and peanuts. *Journal of chromatography A* 2008;1209:271–274.

[96] Wang J, Liu XM. Contamination of aflatoxins in different kinds of foods in China. *Biomedical and environmental sciences* 2007;20:483–487.

[97] Castegnaro M, Tozlovanu M, Wild C, Molinie A, Sylla A, Pfohl-Leszkowicz A. Advantages and drawbacks of immunoaffinity columns in analysis of mycotoxins in food. *Molecular nutrition & food research* 2006;50:480–487.

[98] Trucksess MW, Weaver CM, Oles CJ et al. Determination of aflatoxins B_1, B_2, G_1, and G_2 and ochratoxin A in ginseng and ginger by multitoxin immunoaffinity column cleanup and liquid chromatographic quantitation: collaborative study. *Journal of AOAC International* 2008;91:511–523.

[99] Stroka J, Van Otterdijk R, Anklam E. Immunoaffinity column clean-up prior to thin-layer chromatography for the determination of aflatoxins in various food matrices. *Journal of chromatography A* 2000;904:251–256.

[100] Cheraghali AM, Yazdanpanah H, Doraki N et al. Incidence of aflatoxins in Iran pistachio nuts. *Food and chemical toxicology* 2007;45:812–816.

[101] Fazekas B, Tar A, Kovacs M. Aflatoxin and ochratoxin A content of spices in Hungary. *Food additives and contaminants* 2005;22:856–863.

[102] Arroyo-Manzanares N, Garcia-Campana AM, Gamiz-Gracia L. Multiclass mycotoxin analysis in Silybum marianum by ultra high performance liquid chromatography-tandem mass spectrometry using a procedure based on QuEChERS and dispersive liquid-liquid microextraction. *Journal of chromatography A* 2013b;1282:11–19.

[103] Vishwanath V, Sulyok M, Labuda R, Bicker W, Krska R. Simultaneous determination of 186 fungal and bacterial metabolites in indoor matrices by liquid chromatography/tandem mass spectrometry. *Anal Bioanal Chem* 2009;395:1355–1372.

Index

Bei Fragen zur Produktsicherheit wenden Sie sich bitte an:
If you have any questions regarding product safety,
please contact:

Walter de Gruyter GmbH
Genthiner Straße 13
10785 Berlin
productsafety@degruyterbrill.com